Cover Art

CODE RED FOR PEYTO GLACIER
Oil on canvas, 91x116 cm, 2021.

UN Secretary General António Guterres said in October 2021 that the IPCC Climate Report is a "code red for humanity." Climate scientists have said a catastrophe can be avoided if the world acts fast, and that deep cuts in emissions of greenhouse gases can limit rising temperatures. However, glaciologists estimate that the negligible progress in reducing emissions makes the survival of many mountain glaciers extremely unlikely. This painting shows a conceptual rapidly melting Peyto Glacier in Alberta, Canada, under the intense record heat dome of summer 2021, when the glacier retreated 200 m, roughly ten times its recent rate. The melt has continued at an extraordinary rate in the following three summers. Many mountain rivers draining glaciers in Western Canada sustained abnormally high flows in the heatwave of 2021. Similar events have occurred recently in the Alps, Andes, Hindu Kush, Pamirs, Himalayas and other high mountain ranges around the world.

Published in 2025 by **the United Nations Educational, Scientific and Cultural Organization**,
7, place de Fontenoy, 75352 Paris 07 SP, France

and **the University of Saskatchewan**,
105 Administration Place, Saskatoon, Saskatchewan, S7N, 5A2, Canada

© UNESCO/ University of Saskatchewan, 2025

Printed in Canada

ISBN 978-92-3-100751-4
DOI: 10.54677/QOIR6932

This publication is available in Open Access under the Attribution-ShareAlike 3.0 IGO (CC-BY-SA 3.0 IGO) license (http://creativecommons.org/licenses/by-sa/3.0/igo/). By using the content of this publication, the users accept to be bound by the terms of use of the UNESCO Open Access Repository (https://www.unesco.org/en/open-access/cc-sa).

The designations employed and the presentation of material throughout this publication do not imply the expression of any opinion whatsoever on the part of UNESCO concerning the legal status of any country, territory, city or area or of its authorities, or concerning the delimitation of its frontiers or boundaries.

The ideas and opinions expressed in this publication are those of the authors; they are not necessarily those of UNESCO and do not commit the Organisation.

Global Water Futures (GWF) is a research programme led by the University of Saskatchewan.

Authors: John W. Pomeroy, University of Saskatchewan; Trevor D. Davies, University of East Anglia; Gennadiy V. Ivanov, Norwich Studio Art Gallery
Editorial Commentary: Rita Marteleira, UNESCO; Zoë Johnson, University of Saskatchewan
Review: Anil Mishra, UNESCO; Patrycja Breskvar, UNESCO; Emmanuelle Robert, UNESCO; Yiline Zhao, UNESCO; Hee Eun Ahn, UNESCO
Acknowledgements: Yuqing Luo, UNESCO
Graphic Design: Gennadiy V. Ivanov; Stacey Dumanski, University of Saskatchewan
Cover photo: © Gennadiy V. Ivanov, "Code Red for Peyto Glacier", Oil on Canvas, 91x116cm, 2021 (fragment)
Back cover photo: © John Pomeroy; Ivanov sketching the Saint Elias Icefield at -17 °C
Photographs: All photographs © their respective owners.
Illustrations: © Gennadiy V. Ivanov

SHORT SUMMARY

Uniting Art and Science for Glaciers' Preservation

Glaciers have been the largest contributor from the melting cryosphere to rising sea levels, which are increasing at a rate of about 3 mm per year. If current trends continue, glaciers could contribute an additional 160 mm to global sea level rise by 2100, with Greenland adding 140 mm and Antarctica 100 mm. By that same year, over 400 million people could be at risk of flooding, including residents of some of the world's largest cities.

Facing these concerns, the International Year of Glaciers' Preservation 2025, aims to raise global awareness about the critical role of glaciers, snow and ice in the climate system and the hydrological cycle, and the economic, social and environmental impacts of the impending changes in the Earth's cryosphere. Furthermore, it will be also a contribution towards the International Decade of Action for Cryospheric Sciences, 2025-2034 and the 9th phase of IHP (IHP IX 2022-2029) "Science for a Water Secure World in a Changing Environment".

>400 million people could be at risk of flooding due to sea level rise and glacier melting by 2100

This publication creatively unites science and art, with over 140 original artworks to encourage knowledge exchange and best practices regarding glacier preservation and adaptation strategies.

"Since wars begin in the minds of men and women it is in the minds of men and women that the defences of peace must be constructed"

THE GREAT THAW

A HOMAGE IN ART TO THE VANISHING CRYOSPHERE

JOHN W. POMEROY
TREVOR D. DAVIES
GENNADIY V. IVANOV

Contents

Preface — *i*

Foreword — *iii*

Acknowledgements — *v*

1. Introduction — 1

    1.1 The International Year for Glaciers' Preservation 2025 and the UN Decade of Action for the Cryospheric Sciences, 2025-2034 — 2

    1.2 The Global Water Futures Programme — 5

    1.3 UNESCO International Hydrological Programme — 7

    1.4 The Transitions Art-Science Project — 9

2. Background — 25

    2.1 Climate Change Science and Need for Mitigation — 26

    2.2 Adaptation to the changing cryosphere and its impact on water resources and sustainability — 37

3. Mountains and Forests — 49

    3.1 Mountains — 50

    3.2 Forests — 97

4. Downstream — 111

    4.1 Rivers and Lakes — 113

    4.2 Ecosystems and Agriculture — 125

    4.3 Predictions — 131

    4.4 Challenges and Solutions — 137

5. Conclusions — 141

6. Suggested Reading — 153

References — 155

Authors on the Saint Elias Icefield, Yukon Territory, Canada

Global Water Futures Observatories station at Athabasca Glacier, Alberta, Canada.

Preface

Today, snow and ice play a crucial role in global climate, hydrological, and ecological systems, though they are rapidly disappearing. Although through the geological past they have played greater or lesser roles than today, for humans, snow and ice have been ever-present. Modern humans evolved on an ice-age planet. But still, snow and ice have waxed and waned, with enormous impacts on humanity and our life-support ecosystems. These effects have been indirect, via changing earth systems, and have also been direct, via their seasonal and interannual fluctuations changing habitats, mobility, and sea level. The earliest human agricultural civilisations were formed to manage water for irrigation or flood control, water with its sources in high elevations, many of which were snow and glacier-fed mountain headwaters.

Over the last 12,000 years, and until recent decades, the global climate has been relatively stable, helping human communities to flourish. However, even during this relatively stable period, there have been climatic fluctuations sufficiently large to influence human behaviour. One such fluctuation was the Little Ice Age (1300s to early 1800s), which led to short-term growth in glaciers and snow-cover.

We are now experiencing a new fluctuation, driven by anthropogenic emissions of carbon dioxide, with an opposite effect. Amongst the more visible impacts have been those on the frozen regions of the world called the cryosphere, including ice sheets, sea ice, permanently frozen ground or permafrost, seasonal snowpacks and glaciers. The consequences have been, and will continue to be, dramatic.

The melting of land ice is causing sea level to rise. The thawing of permafrost is destroying ecosystems, changing the lives of Indigenous Peoples, as well as impacting infrastructure. Changes to snow cover are influencing water availability, rivers, lakes, and ecosystems. To many, though, the most obvious impact has been on glaciers.

Mountain glaciers are relatively accessible and have been recorded in art since the seventeenth century and in photography for over a hundred years, during which time scientists also began observing their retreat. Thanks to art, photography, and science, the decline of glaciers since the end of the Little Ice Age has been well characterised.

Glacier melt is now at record fast levels around the world. Global snow covers are declining to all-time lows since satellite observations began in the 1970s, and ancient permafrost is thawing throughout the circumpolar North, exposing viruses, releasing the potent greenhouse gas, methane, and the occasional mammoth from its icy grave. It is now time to prepare for "cryospheric destruction," and it will likely be a distressing experience for most of us. We must prepare for earlier, smaller, and less reliable snowmelt and, hence, less reliable river flows and lake levels that supply drinking water, irrigation water, hydropower, and cold-loving fish. We must prepare for the end of our beautiful mountain glaciers. We must prepare for minimal sea ice in the Arctic Ocean in the summer and sea level rise on all coasts. We must prepare for ecological devastation in millions of lakes as they lose their ice cover, and winter storms load them with sediment and excessive nutrients leading to harmful algal blooms. We must prepare for permafrost thaw, forest collapse, and wildfires burning circumpolar boreal and sub-arctic forests. And we must prepare for food insecurity as the snowmelt that recharges soil moisture and supplies irrigation dwindles and arrives earlier, if at all.

The Transitions Project has been a collaboration between an artist and scientists, who together, sought to fuse art and science to explain the impact of climate change and motivate interest and involvement in the enormous challenges that a warming world poses for human societies. Since its inception in 2019, much of the effort of the Transitions Team has been devoted to portraying the impacts of climate change in cold regions and presenting the underlying science in an accessible way.

The Transitions Art-Science team is proud that UNESCO is publishing some of its work to mark the International Year of Glaciers' Preservation 2025 and the UN Decade of Action for Cryospheric Sciences – 2025-2034. The team also congratulates the UN on such vitally important and extraordinarily timely initiatives.

John W. Pomeroy, Canmore, Canada
Trevor D. Davies, Norwich, England
Gennadiy V. Ivanov, Norwich, England

Foreword

Glaciers play a crucial role in the Earth's hydrological cycle, providing fresh water to approximately 2 billion people (22% of the global population) and helping regulate the global climate by reflecting sunlight and storing carbon. However, over recent decades, global warming has caused widespread shrinkage of the cryosphere, resulting in significant mass loss from ice sheets and glaciers and a reduction in snow cover. These changes have altered the seasonality and quantity of runoff in glacier-fed river basins, leading to diminished water resources, declining agricultural yields, increased water scarcity, and rising global sea levels. A recent study by UNESCO highlights the accelerated melting of glaciers in World Heritage sites, with glaciers in a third of sites set to disappear by 2050.

We know mountains are "water towers" for many regions, and at least half of the world's population depends on water from mountain headwaters. We also know that mountains are among the most sensitive ecosystems, and they are being impacted by climate change faster than other terrestrial habitats, becoming unique indicators of global warming. This will considerably affect these regions' large, often vulnerable populations and livelihoods. UNESCO's Intergovernmental Hydrological Programme (IHP) is a key platform for scientific networking and cooperation to assess and monitor changes in snow, glaciers and water resources and provide adaptation options.

In December 2022, the UN General Assembly declared 2025 as the International Year of Glaciers' Preservation and 21 March as World Day for Glaciers, starting in 2025. The resolution invited UNESCO and WMO to facilitate implementation of the International Year and observance of the World Day. UNESCO, in collaboration with WMO, is spearheading the implementation of the resolution, focusing on raising awareness and implementing strategies for glacier preservation.

The International Year for Glacier Preservation seeks to raise awareness about the critical state of glaciers worldwide, highlighting their essential role in the climate system and the hydrological cycle. This initiative aims to strengthen the connection between the water and climate agendas and draw attention to the cascading economic, social, and environmental impacts of the changes occurring in the Earth's cryosphere.

Awareness raising and education will be central to preserving these vital natural resources, supported by sound scientific research and international collaboration. This effort will encourage nations and organisations to share knowledge and strategies for glacier conservation. The initiative seeks to inspire collective action and responsibility in protecting these ecosystems for future generations by advocating for urgent climate action and promoting sustainable practices.

"The Great Thaw: A Homage in Art to the Vanishing Cryosphere" creatively addresses these pressing challenges by combining art and science to raise awareness about glacier preservation. The book presents a powerful collaboration between artists and scientists, blending evocative artwork with meaningful explanations of cryospheric science. Its goal is to engage and inspire the audience to act in addressing the global issues caused by a warming world, particularly concerning vulnerable communities worldwide.

Lídia Arthur Brito,
Assistant Director-General for Natural Sciences

Acknowledgements

The Authors wish to thank the Canada First Research Excellence Fund and University of Saskatchewan for supporting the Global Water Futures Artist-in-Residence, Gennadiy Ivanov, the GWF Transitions Project and the research that underpins this book.

We thank UNESCO's Intergovernmental Hydrological Programme (UNESCO-IHP) and Global Water Futures (GWF) for contributing to the costs of publishing this book.

The contributions of many people made this book possible – we thank Prof. Sean Carey of McMaster University, Prof. Phillip Marsh of Wilfred Laurier University, Prof. Rich Petrone of University of Waterloo, and Prof. Cherie Westbrook, Dr. Warren Helgason and Dr. Graham Strickert of the University of Saskatchewan for access to their field sites, the assistance of their field research teams and descriptions of their research. The researchers of the University of Saskatchewan Coldwater Laboratory in Canmore, Canada made exceptional efforts to accommodate the needs of the Transitions Project.

We also thank the Centre for Hydrology, University of Saskatchewan for the assistance of Ms. Zoë Johnson and Ms. Stacey Dumanski in editing this book, Ms. Joni Onclin in logistics and organisation and Dr. Chris DeBeer, Prof. Timothy J. Osborn, and Dr. Andre Bertoncini for map and figures.

1. Introduction

1.1 The International Year for Glaciers' Preservation 2025 and the UN Decade of Action for the Cryospheric Sciences, 2025-2034

In December 2022, the UN General assembly adopted the resolution to declare 2025 as the International Year of Glaciers' Preservation (IYGP), accompanied by the proclamation on the 21st of March of each year as the World Day for Glaciers, starting in 2025. The International Year and World Day for Glaciers aim to raise global awareness about the critical role of glaciers, snow and ice in the climate system and the hydrological cycle, and the economic, social and environmental impacts of the impending changes in the Earth's cryosphere. In this regard, and in addressing the issues related to accelerated melting of glaciers and its consequences, the International Year and World Day for Glaciers also aim to encourage the exchange of knowledge and best practices regarding glacier preservation and adaptation strategies.

The melting of glaciers and snowpacks and the thaw of permafrost affects everyone. People living in coastal areas are affected by sea level rise, people living in high mountain areas are more prone to the risks of flooding, landslides, and avalanches, and people living in downstream areas are vulnerable to changes in water supply from snow and glaciers. Melting glaciers and ice sheets were identified as one of the largest contributors of sea level rise in the past decades according to the Intergovernmental Panel on Climate Change (IPCC) in 2022. Glaciers in UNESCO World Heritage sites are melting at an alarming rate, with glaciers in a third of the sites set to disappear by 2050. In the same time frame, most remaining tropical glaciers in South America, Africa and Asia will vanish. Glacier loss is also accompanied by the loss of biodiversity, especially of endemic species, as well as the loss of cultural values and traditional ways of life.

Glacier and snow cover retreat clearly present a serious threat to natural and human water supplies in many parts of the world. The IYGP and World Day for Glaciers are therefore focusing on providing concrete recommendations to address climate change impacts on the cryosphere, advocating for more ambitious mitigation, convening countries and communities affected by glacier loss for sharing knowledge and best practices for preservation and adaptation, and raising international funding for adaptation action in affected areas.

The objectives of the IYGP are to:

Raise Awareness: Increase public and stakeholder awareness at all levels about the importance of glaciers in the climate system, hydrological cycle, and global water resources, the differential impacts of glacier changes on downstream communities and ecosystems, and the urgent action needed to develop adaptation strategies.

Promote Action: Facilitate the implementation of sustainable measures and best practices for the preservation of glaciers, encouraging transboundary cooperation, knowledge-sharing, and innovative approaches, including encouraging behavioural change through culture and the arts.

Enhance Scientific Understanding: Support scientific research and monitoring initiatives to improve the understanding of glacier changes, the impacts of climate change, possible loss and damages, and the implications for communities, ecosystems, and water resources.

Strengthen Policy Frameworks: Advocate for robust policy frameworks at national, regional, and international levels to address the preservation of glaciers, incorporating climate change adaptation, sustainable water management, and disaster risk reduction strategies.

Strengthen Financial Support: Ensure financial resources are made available to support glacier monitoring and management of the impact of climate change on glacier melt and downstream impacts.

IYGP 2025 offers a unique opportunity to prioritize the preservation of these vital mountain, polar and downstream ecosystems. By engaging stakeholders, raising awareness, promoting action, and strengthening policy frameworks, we can work together to safeguard glaciers and the cryosphere and the invaluable services they provide to humanity and the environment. The IYGP, facilitated by UNESCO and WMO, in collaboration with governments and relevant organizations, will be a key steppingstone to ensure a sustainable future for glaciers and the communities that depend on them.

www.un-glaciers.org

Introduction

In August 2024, the UN General Assembly adopted a new resolution, the "Decade of Action for Cryospheric Sciences, 2025–2034". The Assembly proclaimed the Decade to address the challenges associated with melting glaciers and changes to the cryosphere by advancing related scientific research and monitoring. Introducing that text, the representative of France, speaking also on behalf of Tajikistan, spotlighted the vulnerability of glaciers and the poles to climate change and their role in regulating climate and ocean levels, and preserving biodiversity. "This UN Decade will provide a political impetus needed to make this issue a priority on the multilateral agenda," she said. UNESCO is invited to lead the Decade of Action, in collaboration with other relevant organizations of the United Nations system as well as other stakeholders. The Decade will link the IYGP to the International Polar Year (2032-2033) and provide a sustained scientific focus on snow and ice to help in understanding and designing adaptation and mitigation measures to one of the most devastating and prominent impacts of a warming climate – the end of the Ice Age that modern humanity has evolved in.

Athabasca Glacier and Columbia Icefield, Jasper National Park, Canada is a research basin for Global Water Futures Observatories.

© Gennadiy Ivanov

1.2 The Global Water Futures Programme

Our water is at risk. Globally, we are facing unprecedented water-related challenges, with half of the world's population dependent on water from cold regions. Facing these challenges, water scientists in Canada and globally asked the question, *"how can we best forecast, prepare for and manage water futures in the face of dramatically increasing risks?"*.

Global Water Futures (GWF), the largest university-led freshwater science research project in the world, is a Canadian funded programme, led by the University of Saskatchewan, Wilfrid Laurier University, University of Waterloo, and McMaster University. It was formed in 2016 to conduct research to better understand water challenges and find sustainable solutions to some of the world's most complex water problems in an era of climate heating. With a budget of more than $100 million (USD), GWF made key developments in transdisciplinary science, innovative environmental monitoring systems, predictive modelling tools and novel user-focused approaches to finding solutions for freshwater sustainability. These achievements have put new knowledge into action, provided risk management technologies, decision-making tools and have provided other evidence-based solutions to water challenges in Canada and around the world. To do this GWF scientists designed a three-part mission to **1) improve disaster warning; 2) predict water futures;** and **3) inform adaptation to change and risk management.** These three pillars were created following extensive discussions with hundreds of user groups across the country and around the world, which informed specific research priorities that diagnose and predict change in cold regions, develop big data and decision-support systems, and design user solutions. GWF supported this mission through advocacy, influence in policy and practice, and research projects designed to produce meaningful solutions and tools.

Over the last eight years, GWF has improved disaster warning using observations from 76 instrumented research basins across Canada and organized 38 instrumented basins in high mountains around the world. It has developed the first flood forecasting system for sub-Arctic and Arctic Canada and novel automated monitoring of water quality in large lakes. It has improved the prediction of water futures by developing the first national water model for Canada and applying it to over 5 million km² across Canada, to mountain river basins in China, Nepal, Iran, India, Chile, Spain, Germany, Austria, USA, and to mountain headwater basins on every continent. GWF developed and applied climate, cryosphere and hydrology models that informed adaptation to climate and development changes and managing risk from future floods and droughts. It has strengthened national resiliency by proposing the creation of a Canada Water Agency to coordinate Canadian water science and policy functions.

Working with practitioners and knowledge users, GWF scientists have improved approaches that detect invasive species and algae in water bodies and viruses like SARS-CoV-2 in wastewater. The programme has codeveloped solutions that improved resilience and helped achieve water sustainability with Indigenous communities and sectoral users across Canada, with advances in addressing the impacts of wildfire on northern communities and the supply and quality of drinking water. GWF researchers have also provided evidence for managing agriculture to improve lake and wetland health whilst improving food productivity. They have calculated the national and global loss of glaciers, snowpacks and permafrost and the impacts on ecosystems and water resources. These achievements and others helped spur the International Year of Glaciers' Preservation 2025, and the Decade of Action for Cryospheric Sciences 2025-2034, and are what made the Transitions Art-Science project possible.

Hydrometeorological station in Fortress Mountain Research Basin, Alberta, Canada, and part of the Global Water Futures Observatories.

Through collaborations with artists, researchers, and Indigenous Peoples, GWF has supported activities that use visual art, storytelling, and multimedia to translate scientific findings into accessible formats. The Transitions art-science collaboration developed hundreds of scientifically described artistic creations on cold regions that were exhibited in Canada and England. The Virtual Water Gallery (VWG; https://www.virtualwatergallery.ca/), an online art-science project funded by GWF, fostered collaborations between artists and scientists from across Canada, including Indigenous artists and knowledge holders, to share water perspectives to find creative and holistic solutions to our water challenges. Launched in Summer 2020, the VWG started as an online art-science gallery and eventually went live with in-person exhibitions in 2022 and 2023. Through attendee surveys at the in-person exhibitions, the VWG has validated the significance of art in changing knowledge levels and intended behaviours related to water-related climate change mitigation.

1.3 UNESCO International Hydrological Programme (IHP): From Science to Solutions

Celebrating in 2025 its 60th anniversary of work in the water sciences realm, UNESCO has been supporting glaciologists, hydrologists and water resources specialists from all over the world and providing them with a platform to undertake research on the cryospheric changes and impacts on water resources. The need for a worldwide inventory of existing perennial ice and snow masses was first considered during the International Hydrological Decade, declared by UNESCO for 1965-1974. However, there remain clear gaps in monitoring of the glacier system in all mountainous regions.

The ninth phase of UNESCO's Intergovernmental Hydrological Programme (IHP-IX, 2022-2029) puts science to action for a Water Secure World in a Changing Environment, and it aims to undertake a comprehensive assessment of the state of the snow, glacier and permafrost impacted by climate change. The cryosphere has been identified as a priority in the IHP-IX key expected outcomes, which emphasized the need for a comprehensive cryosphere assessment and to develop and sharing of methods to monitor changes in the cryosphere system (snow, glacier, and permafrost), runoff formation from melting glaciers erosion and sediment transport.

IHP will likewise promote scientific research to enhance understanding of water availability from the cryosphere and in aquifers and call for science-based policy decisions and strengthen the adaptation capacity of countries to climate change impacts through assessment, promotion of regional cooperation, and stakeholder engagement. IHP IX will support strengthening national and regional capacities to monitor Glacial Lake Outburst Flood (GLOF) hazards and the development of early warning in response to climate change for several regions will be implemented.

Furthermore, around 18,600 glaciers have been identified in 50 World Heritage sites. These glaciers span an area of about 66,000 km, representing almost 10% of the Earth's glacierized area. Research studies performed with satellite data highlight that these glaciers have been retreating at an accelerating rate since 2000. Projections indicate that glaciers in one-third of World Heritage glacierized sites will disappear by 2050 regardless of the applied climate scenario and glaciers in around half of all sites could almost entirely disappear by 2100 in a business-as-usual emissions scenario.

In December 2022, the UN General Assembly declared 2025 as the International Year of Glaciers' Preservation (IYGP 2025) and 21 March as World Day for Glaciers, starting in 2025. The resolution invited UNESCO and WMO to facilitate implementation of the International Year and observance of the World Day. UNESCO, in collaboration with WMO, is spearheading the implementation of the resolution, focusing on raising awareness and implementing strategies for glacier preservation.

In addition, the United Nations General Assembly's declaration of the Decade of Action for Cryospheric Sciences (2025-2034) represents a significant milestone in advancing international scientific cooperation. The resolution recognizes the vital role of the cryosphere in regulating Earth's climate system and ensuring freshwater availability. As a lead agency for the implementation of the Cryosphere Decade, UNESCO-IHP will coordinate with Member States, scientific communities, and relevant stakeholders to strengthen cryosphere research and monitoring capabilities, while developing comprehensive strategies for adaptation.

This publication "The Great Thaw: A Homage in Art to the Vanishing Cryosphere" is a contribution to the IYGP-2025, International Decade of Action for Cryospheric Sciences (2025-2034) and to the implementation of the 9th phase of IHP (2022-2029).

1.4 The Transitions Art-Science Project

Transitions was an interdisciplinary project sponsored by the Global Water Futures programme, in which the UK-Canadian Transitions team worked to bring art and science as close together as possible to highlight the impacts of climate change. Artist Gennadiy Ivanov learned something of the science behind climate change and has joined scientists on research visits and measurement campaigns in remote parts of Canada and around the world.

Ivanov produced his field paintings very rapidly, usually within 15-20 minutes. This was a helpful attribute for days when it was very cold, a blizzard was about to blow, or the helicopter pilot was starting to point to his watch. Most paintings were produced in discussion with the scientists as they went about their work, as Ivanov witnessed the observational techniques deployed in the field, or in conversation about the physical explanations of the features being painted.

Working beside Ivanov, scientists Trevor Davies and John Pomeroy learned to appreciate the power of art, and how Ivanov's skills and techniques could be used to emphasise the most salient physical characteristics of landscapes and natural phenomena. This two-way interaction was, perhaps, most rewarding when Ivanov produced his studio paintings based on his recollections, photographs, and field paintings. Sometimes, paintings went through several iterations to find the best combination of representation and interpretation to hit the two buttons: artistic impact, and scientific coherence and message.

Some of Ivanov's studio paintings were not of real locations. When he let his imagination run riot, he produced paintings which conjured an impression of what he had seen and heard on the field campaigns. Ivanov calls these his "evocative paintings".

Ivanov sketching by the hydrometeorological station at Fortress Mountain.

Artist: Gennadiy V. Ivanov

Gennadiy Ivanov is a UK-based artist. Originally from Belarus, and born in Russia, he graduated with an MA in Fine Art from Norwich University of the Arts. He uses synthetic thinking and a wide range of techniques. This allows him to work simultaneously in several directions and styles. His paintings demand intellectual, as well as visual and emotional, effort from the viewer.

Ivanov works in a wide variety of media, including paintings, installations, drawings, and photography. He attempts to bring these media together in a provocative collision. Ivanov's work mines a territory not only defined by the viewer's understanding of recent world political history, but expands into the flowing social veins of an underground style; this can embrace cultural and creative fields such as street art, fashion, and experimental dance and music.

"I had painted scenes of the Norfolk coastline in the knowledge that coastal erosion is a challenge for the County, and that rising sea-level resulting from climate change is an increasing problem. The predictions for future sea-level rise alarmed me, and I wanted to capture something of the existing beauty of our coast and, somehow, represent this growing challenge. The experience of producing that series of paintings led me to contact Professor Trevor Davies. I knew that Norwich was a hotbed of climate and environmental research and the possibility of collaboration with a local scientist was something I felt I had to explore. I know that art can stimulate interest in, and raise awareness of, issues which deserve more attention; and I hoped to be able to make a contribution in this regard. What has happened since I contacted Professor Davies has astounded me. I have unbounded admiration for the dedication and sheer hard work of the scientists. I feel privileged to have been able to witness their important research involving sophisticated instrumentation in remote and difficult terrain, some locations of which could only be reached by specialised transport. My visits were possible only with the fantastic support of the Global Water Futures research programme and, especially, its Director, Professor John Pomeroy, and they opened my eyes. I hope my paintings will do justice to the scientists, and adequately capture a sense of the challenge that is climate change. Most importantly I hope they do justice to the glaciers".

Scientist: Trevor D. Davies

Trevor Davies's research interests include climate, snowmelt processes, air pollution, and chemical hydrology. He has been Director of the Climatic Research Unit, University of East Anglia (UEA), UK and a founding member of the Tyndall Centre for Climate Change Research, which is headquartered at the same university. He established the Fudan Tyndall Centre at Fudan University in Shanghai, where he was a Distinguished Professor. He headed the School of Environmental Sciences at UEA, where he also took the position of Pro Vice-Chancellor for Research, Enterprise and Engagement.

"I was struck by Gennadiy's paintings of the Norfolk coastline and felt that his style and imagination could work very well within a wider collaboration with scientists. If this was to be a genuine fusion of art and science, I thought we needed to understand each other's methods and techniques, and the processes which go into producing our outcomes; and we needed to travel together along the route to the final representation on paper or canvas.

"Gennadiy was a willing student of the fundamentals of climate science. One of the aspects of his work which attracted me was its often fluid quality. Water plays an important role in climate: transitions between the phases of water, and the distribution of those phases, influences the energetics of the climate system. Water is also a vector of impact: too much of it (floods); or too little (droughts, fires).

"To cope with change, we will also have to make transitions in the way we do things. Transitions seemed to be a good title for this project. I thought it would be beneficial for Gennadiy if he could experience and portray landscapes which were less gentle than those of Norfolk and where some of the impacts of climate change are dramatic. In particular, there was an opportunity for him to paint the mountain glaciers which are under great stress because of climate change. I contacted a long-standing research colleague, Professor John Pomeroy, whom I first worked with when he did post-doctoral research with me at UEA. John is Director of the crucially important Global Water Futures programme headquartered in Saskatchewan. Canada is experiencing an increasing frequency of extreme events, directly related to pronounced climate change; unprecedented floods, droughts, vegetation fires, permafrost melting, and snow- and ice-cover reductions. These dramatic changes are having significant impacts on big-city infrastructure, Indigenous Peoples, agriculture, and hydro-electricity generation. I thought Gennadiy could project some visual emotion from these impacts".

Scientist: John W. Pomeroy

John Pomeroy's research interests include snow and ice hydrology, cold regions climate change, water quality, water predictions and observation. He is a Distinguished Professor in the Department of Geography and Planning at the University of Saskatchewan, a Fellow of the Royal Society of Canada, UNESCO Chair in Mountain Water Sustainability, Canada Research Chair in Water Resources and Climate Change, Director of the Centre for Hydrology, Director of the Global Water Futures Programme, and Co-Chair of the UN Advisory Board for the International Year for Glaciers' Preservation 2025.

© Marilyn Pomeroy

"When Professor Trevor Davies sent me examples of Gennadiy's paintings of Norfolk, I felt that his style and approach could help visualise aspects of climate change and some of the Global Water Futures research related to human-induced changes in our freshwater environment. I was pleased to be able to invite him to Canada to witness some of our research and the pronounced impacts of climate change on the cryosphere.

"Trevor and I have a long-standing collaboration on field research in Canada and the UK, so it was an appropriate partnership. An important element of the Global Water Futures (GWF) programme is public engagement and awareness-raising. We have worked hard to co-develop our research with users including Indigenous Peoples, and so I was keen that Gennadiy join our routine checking of observation stations and see examples of the impacts of climate change where remote communities and Indigenous Peoples are being most affected. These trips involved intense discussions of the science outcomes and changes we were observing. As scientists, we were impressed with how quickly and effectively Gennadiy captured his first impressions of what he was seeing, sometimes under harsh conditions. It was important that we could come to a joint understanding of what was being portrayed as he produced his final oil paintings. We saw this as an iterative and collaborative process, which has been very successful. Science colleagues in Global Water Futures (GWF) were thrilled by the art that Gennadiy has produced, and this has inspired development of the GWF Virtual Water Gallery https://www.virtualwatergallery.ca/ where 28 artists and scientists, including Indigenous artists, came together to co-produce art that interprets GWF science and community knowledge. Further, this art has been instrumental in visualising the case for the International Year of Glaciers' Preservation and the coupled water and climate crisis at various UN and COP meetings and I think has positively impacted international deliberations to find solutions for this crisis".

THERE ARE STILL WOLVES

Field drawing, pastel on paper, 30x42 cm, 2019.

The ecosystem at Wolf Creek is changing rapidly. Here are a number of scientific monitoring stations which have observed the environmental change over the last 25 years. The vegetation is responding. Shrubs are growing more prolifically. Wolves still roam, but other predators, including lynx and coyote, are moving northwards. The ecosystem will continue to change.

Hydrometeorological station at Wolf Creek Research Basin, Yukon Territory, Canada, and part of the Global Water Futures Observatories.

THE RED INVASION AT WOLF CREEK
Oil on canvas, 91x116 cm, 2019.

The Global Water Futures Observatories station was established in a valley covered with moderate height tundra shrubs in the 1990s. Since then, shrubs have grown in height and are now above the instruments and overwhelming the station. The red bushes are represented as flickering flames consuming the instrumentation, heating the surroundings. They carry black seeds of the continuing invasion. This painting is a good example of iterative collaboration between artist and scientists, who felt that early versions of the impression did not adequately represent the true scale of the invasion and disruption to the instrumentation. The station is no longer useful as a source of areally representative observations but is retained to record the modifications caused by the vegetation close to the ground.

"I feel that this painting truly captures something of - and from - the atmosphere of this beautiful location."
- Ivanov

HAIL ON KATHLEEN LAKE
Field drawing, pastel on paper, with a contribution from a hailstorm, 30x42 cm, 2019.

Kathleen Lake, in Kluane National Park, Yukon, Canada, is still ice covered. This scene was captured when the artist was on the lake, together with two Indigenous People who were fishing through a hole in the 80 cm thick ice. An enormous hailstorm started, and the stippled appearance of the drawing is the result of hailstones falling on the paper.

Kokanee salmon live in the lake; a landlocked version of Sockeye salmon, trapped from returning to the ocean by surging glaciers in previous ice advances, which have adapted to living wholly in freshwater. The Kokanee population is now recovering from a crash with a low point in 2008. The reasons are not fully understood, but it may be related to an interaction between climate and the hydrological cycle (including hydro-chemistry); an example of the importance of having good scientific observations.

Thawing permafrost at Caribou Creek, Northwest Territories, Canada.

CARIBOU CREEK THAW SLUMPS
Field painting, mud, gouache on paper, 30x42 cm, 18x24 cm, 2019.

According to researchers, as global temperatures rise, so too has the amount of permafrost thaw contributing to massive ice landslides called retrogressive thaw slumps. Here, ice-rich permafrost is thawing and slumping along the banks of Caribou Creek, causing forest collapse. GWF models suggest that 90% of permafrost in the Mackenzie River Basin which encompasses Caribou Creek, will thaw by the end of this century under business as usual climate change trajectories.

BREAKFAST WITH SCIENTISTS

Oil on canvas, 150x120 cm, 2019.

A vital part of the Transitions climate-art project is discussion with the scientists, not only in the field but also in reflection. This conceptualization of a breakfast conversation with John Pomeroy (left) and Trevor Davies (right) occurred the morning after an exhausting day on the Peyto Glacier in August 2019. On the table is an accumulation of cryoconite; a strange material which consists of ash and soot from wildfires and air pollution, dust, bacteria, fungi, algae and other organisms. It collects on the surface of the glacier and has been increasing over the years as more frequent and more extensive wildfires deposit soot which feeds the algae and microbes, darkens the glacier and contributes to increasing melt rates. Summer melt washes some of it off the glacier surface, and it accumulates in weird formations below the snout of the glacier. Global Water Futures researchers have examined the composition of cryoconite using scanning electron microscopy and DNA sequencing and have shown that the darkening of glaciers from its accumulation has accelerated melt rates by up to 10%.

THE FORTRESS NOW
Pastel on paper, 45x65 cm, 2019.

The Fortress is a mountain with an iconic location in the Canadian Rockies of Alberta, Canada. Fortress Mountain Research Basin in the Canadian Rockies Hydrological Observatory, is part of the Global Water Futures Observatories network of 64 instrumented research basins across Canada that inform the development and testing of water prediction models, monitor changes in water resources, help design solutions to ensure water sustainability and underpin the diagnosis of risks to water security under a rapidly changing climate.

FORTRESS MOUNTAIN AT THE END OF SUMMER
Pastel on paper, 45x65 cm, 2019.

Despite the snowy winter, soil moisture reserves that were replenished by spring snowmelt reached the wilting point by late summer. Groundwater is insufficient to supply adequate moisture for plant transpiration and wetland evaporation in the longer drier, hotter summers that have occurred recently. Alpine vegetation has adapted to the short summer and as it moves into an early senescence after a hot dry summer, adopting brilliant colours that bring exceptional beauty to the high mountain tundra. There are signs that the vegetation is facing increasing drought stress as the summer heat and snow-free period increase.

THE FORTRESS MOUNTAIN
Oil on canvas, 46x25 cm, 2019.

Ivanov uses his field paintings, photographs and videos to produce studio oil paintings. The Fortress is a recreational site for skiing, but also an important and heavily-instrumented scientific laboratory where Global Water Futures scientists study water and energy cycles, including how snow is redistributed by wind and gravity, sublimated, intercepted by forest canopies in the research basin, and melts to form streamflow and replenish soil moisture and groundwater flow.

"Perhaps ironically, most of my other paintings of Peyto, despite its sad state, are amongst the most brightly-coloured I have produced. They were based on a brilliantly-clear blue sky day summer visit in August. On this particular occasion, the glacier was still mostly hidden beneath snow-cover. It was a miserable day; better fitting the emotions which I now feel about this departing feature of the dramatic mountain landscape. I have transposed my darker emotions into this drawing." - Ivanov

GLACIER DECLINE
Charcoal and ink on paper, 170x100 cm, 2021.

Ivanov sometimes used charcoal and ink to help build the foundations for his studio oils. This drawing was based on field sketches and photographs on a particularly cold and grey April day on the Peyto Glacier. In the distance, the glacier is covered by snow. In the foreground are strange deposits of black cryoconite. The cryoconite accumulates on the ice surface and, each year, is washed off by the copious summer meltwater to form mini-mountain ranges, a metre or so high, beyond the glacier snout.

Peyto Glacier, Rocky Mountains, Alberta, Canada, and part of Global Water Futures Observatories.

© John Pom

FORMER PEYTO GLACIER
Pastel on paper, 45x65 cm, 2019.

The crisscross patterns of crevasses and melt channels on the glacier give a sense of the collapse of the ice mass. Much of the foreground detail in this painting was hidden by snow on the artist's first visit to Peyto in April 2019. The landscape was then frozen. In August 2019 the deposits of glacial silt and black cryoconite accumulations - surrounded by, and saturated with, water – were next-to-impossible to walk over. They sucked Davies down up to his knees.

> "A colourful painting; when I look at my own painting as a spectator, the bare moraines and sediments left by the retreating ice give me a sense of destruction, darkness and decay borne in the rapid deglaciation initiated by human-caused climate change. It is a disturbing to task to try to represent the sense of decay during an azure day which produced vivid contrasts and colourations". - Ivanov

MELTING GLACIER
Oil on canvas, 24x30 cm, 2020.

This is one of a series of "evocative paintings" which Ivanov has produced. They are not tied to one location, but come from his mind's eye seeing his reflections, recollections, and moods when he remembers his field excursions with the scientists, and considers the import of what their observations, analysis and predictions mean.

2. Background

Hydrometeorological station in Fortress Mountain Research Basin, Rocky Mountains, Alberta, Canada, and part of the Global Water Futures Observatories.

2.1 Climate Change Science and Need for Mitigation

Climate Change Imperils Glaciers

Most glaciers have retreated since the second half of the 19th century; something which has likely been unprecedented for at least the last 2000 years. Their retreats have been particularly pronounced since the 1990s and, in the last decade or so, they have been at a catastrophic rate.

Given this retreat, the International Year for Glaciers' Preservation 2025 is an ambitious initiative. Local cryo-engineering technological fixes, such as covering or insulating glaciers, are not viable for large scale global deglaciation. The only way to slow and stop the damage to glaciers – and the rest of the cryosphere – and keep within "adaptable" levels is with substantial and rapid cuts in anthropogenic emissions of greenhouse gases, especially carbon dioxide. A commonly-cited target to avoid the most damaging effects of climate change – agreed by international treaty in Paris in 2015 – is to limit global heating to 1.5 °C above the pre-industrial temperature level, as a 20-year average. This is a massive challenge.

The World Meteorological Organisation announced in January 2025 that the year 2024 was the Earth's warmest year on record, with temperatures at about 1.55 °C above the pre-industrial level for the first time. The sort of action required to remain below the 1.5 °C 20-year average target is a 40% reduction in fossil fuel emissions by 2030; then continuing reductions to net zero by the middle of the century; and then "negative carbon", requiring measures to absorb more carbon from the atmosphere than is emitted to it, to 2100.

The UN's Intergovernmental Panel on Climate Change (IPCC) refers to this as the "sustainable track." This track will be tough to follow. The sustainable track is not enough to sustain most glaciers and the planet's current seasonal snow cover; we need to do better. The track we are currently on – the "business-as-usual track" – will lead to a global temperature increase of around 4.4 °C (range 3.3 °C - 5.7 °C) by 2100. The IPCC judges that that our likeliest track to 2100 - assuming we will make considerable reductions in emissions - will lead to a global temperature increase of around 2.7 °C (2.1 °C - 3.5 °C). The likeliest track is, clearly, not good enough for glaciers and snowpacks.

The World of Snow and Ice – The Cryosphere

Mountain glaciers are a relatively small repository of frozen water and hold around 158,000 km³ of ice, compared to: 3,062 km³ of ice held in the peak seasonal snow cover outside of mountains in the Northern Hemisphere, plus 200-500,000 km³ of ice in permafrost, plus 2.9 million km³ in the Greenland icesheet, and 330 million km³ in Antarctica. Moreover, as a source of river discharge, the melt of the seasonal snowpack is usually far more important than glacier ice melt. River and lake ice are critically important to peak flood stage, aquatic ecosystems and transportation in cold regions. The various manifestations of ice in the cryosphere are all important for the function of the earth system, as they are integral parts of our coupled climate and hydrological system. Continued reduction in the occurrence and coverage of glaciers, icesheets, permafrost, and snowpacks will darken and warm the Earth's surface and further accelerate global heating.

It is understandable that the UN has chosen glaciers out of all its components to highlight the threat to the cryosphere. Glaciers are visible to a greater number of people and are a key indicator of the extent of climate change that has occurred. They are, most obviously to more people, in rapid retreat: in peoples' direct experience; from instrumental records; from photographs, drawings and paintings. For centuries they have been understood to be an important source of freshwater for large areas downstream of high mountains. They have great cultural value in mountain societies.

Other components of the cryosphere also have a strong cultural value. Although less visible to large numbers of people they, too, are under serious threat. Permafrost, or permanently frozen ground, has influenced the culture and shaped the working life of Indigenous Peoples in the sparsely populated, but vast tundra and boreal forest regions. In the words of a Chief of the Gwich'in people in the Yukon Territory and Northwest Territories, Canada, it is like "watching a nuclear explosion in slow motion" to observe the thaw of permafrost in his traditional territory. Hundreds of square metres of land have slumped before his eyes, trees have lost their purchase and then toppled over completely. Formerly ice-underlain landscapes are collapsing. Large quantities of "locked-up" carbon dioxide and methane are being released to the atmosphere. The culture and memories of the Gwich'in people are threatened by the rapidly changing landscape and collapsing ecosystems.

Arctic Sea ice has been shrinking by about 12% per decade since the late 1970s. The Greenland and Antarctic ice sheets have both lost mass since the 1990s. In both cases, the melting represents a tiny amount of total ice mass, but enough to contribute around 13.5 mm and 7.4 mm, respectively, to sea level rise. In comparison, glacier melt has contributed around 30 mm to sea level rise since the 1960s. Arctic sea ice melt makes no contribution to sea level rise.

The Northern Hemisphere cryosphere has been hard hit by global heating; Arctic regions are experiencing heating rates around five times higher than the global average. This pronounced heating is because of a strong positive "feedback" effect. Ice and snow cover reflect more of the Sun's radiation than does open water or bare land. North American snow mass has declined by about 10% since 1980 and global snow cover has dropped by about 15% since the early 20th century. This reduction in reflectivity means a greater absorption of solar radiation, further increasing the regional temperature.

The Antarctic Peninsula has warmed by around 3.2 °C since the 1950s; West Antarctica warmed by around 0.1 °C per decade since the 1950s, but experienced cooling from the 1990s onwards. East Antarctica is colder and more stable and has been experiencing slight cooling. Extreme events are starting to affect Antarctica; in March 2022, temperature in central Antarctica climbed rapidly from −55 °C to −10 °C.

John Pomeroy and Greta Thunberg on Athabasca Glacier, Rocky Mountains, Alberta, Canada

The Climate has Always Varied

The global climate had been relatively stable over the last 12,000 years, during which time human communities and civilisations have flourished. But even this relative stability compared to earlier climates had included sufficient variability to greatly impact human societies; an example is the "Little Ice Age" from the 1300s to the early 1800s. The impacts of recent dislocations to climate caused by global heating gases have put our vulnerability, despite our technological advances, into very sharp focus.

There have been major swings in the Earth's climate as we go back through geological time. The planet was in a deep freeze at least twice between 750 and 600 million years ago. At other times, it was essentially free of ice, as at around 50 million years ago. Information from so-called physical proxies such as ocean sediments, ice cores and tree rings can be used to reconstruct a global temperature record, such as that shown in Figure 2.1.1. which covers the period from 66.5 million years ago to 1850.

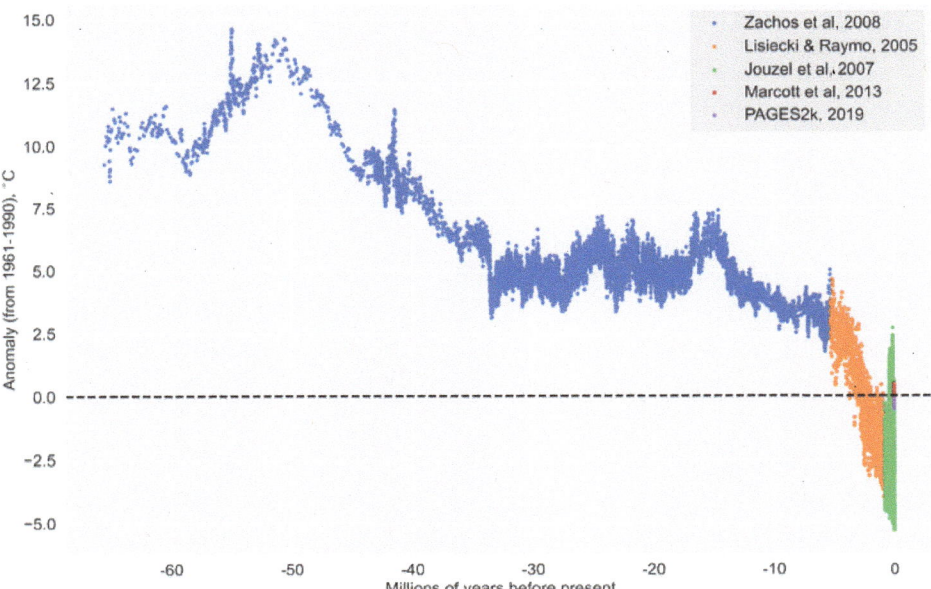

Figure 2.1.1. Global temperature plotted as an anomaly (difference) from the 1961-1990 reference period. *There is a large temperature range; from -5.0 °C to +15 °C. This compares to a range of just -0.5 °C to +1.0 °C from 1850 to the present. Figure courtesy © Timothy J. Osborn, Climatic Research Unit, University of East Anglia, UK.*

From 1850 onwards, instrumental observations were plentiful enough to construct a global temperature record from direct measurements; this is shown in Figure 2.1.2.

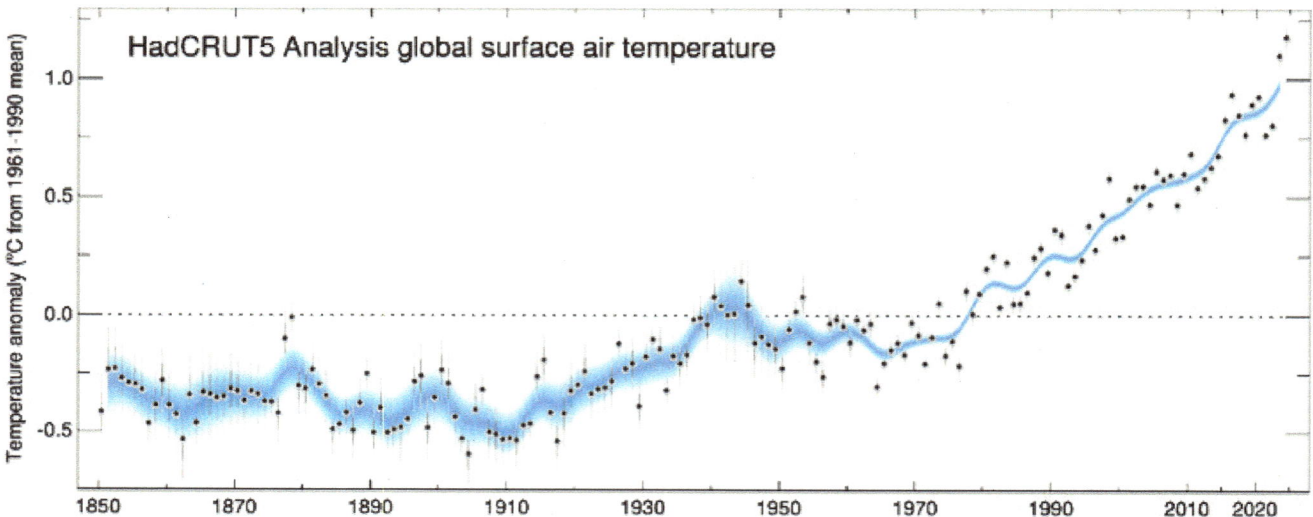

Figure 2.1.2. Global temperature from instrumental observations from 1850 to present. The values are anomalies from the 1961 - 1990 reference period. Figure courtesy © Timothy J. Osborn, Climatic Research Unit, University of East Anglia, UK.

It is important to note that both figures show deviations from a 1961 - 1990 reference period. The 1.5 °C threshold and future temperatures discussed earlier were increases from pre-industrial levels.

Detailed comparisons of modelled future temperatures with the temperature record from proxy data show that, if we manage to take IPCC's likeliest track to 2100, we will experience temperatures last experienced on Earth 2-3 million years ago. If we continue our current business-as-usual track, the Earth will see temperatures last experienced several million more years ago.

The rate of heating caused by greenhouse gases is important. Over the last century, the global temperature has increased at about ten times faster than the average rate of warming since the Ice Age. When global warming has occurred in the past 2 million years, it took about 5000 years to warm by 5 °C. Likely rates of heating over the next century are around 20 times faster. The earth system is responding to very unusual rates of heating. This is why climate scientists warn about "tipping points". A different type of tipping point, which is imperative – given the amount and rate of global heating we are facing - but which we are yet to see, is in human behaviour.

Melting Ice Endangers People Around the World

Glaciers have, up to now, made the biggest contribution from the melting cryosphere to rising sea level, which is increasing at the rate of about 3 mm per year. If we continue our current business-as-usual track, glaciers may contribute another 160 mm to global sea level rise by 2100, with Greenland adding 140 mm, and Antarctica another 100 mm. Thermal expansion of the oceans may add another 310 mm. If we follow our likeliest future track, glaciers will contribute around 110 mm to sea level rise by 2100, Greenland 90 mm, and Antarctica 110 mm, with thermal expansion adding 200 mm. Following the sustainable track might limit glaciers to around 70 mm contribution, Greenland to 50 mm, Antarctica 110 mm, and thermal expansion at 120 mm. Sea level rise, especially when combined an expected increase in storm intensity, will increase coastal erosion. Potentially, more than 400 million people could be at risk from flooding by 2100, including in some of the world's largest cities.

Snowmelt experiment on a frozen lake.

WOLF CREEK RESEARCH BASIN TALL TOWER

Oil on canvas, 100x80 cm, 2023.

To develop predictive models, it is necessary to monitor energy and water balances in representative landscapes across large regions. Forests cover vast areas in the boreal regions. The measurements also include soil moisture; these are invaluable in helping to evaluate wildfire risk.

Wolf Creek Research Basin, Forest Meteorological Tower, Whitehorse, Yukon, Canada, and part of Global Water Futures Observatories.

THE FORTRESS MOUNT NOW
Oil on canvas, 130x170 cm, 2019.

Fortress Mountain Research Basin is an iconic location in the Rockies in Alberta. Its dramatic countenance has appeared in many Hollywood films. It is an important Global Water Futures observation site which automatically records atmospheric, snow and soil conditions. These observations are part of the Global Water Futures Observatories network, which monitors changes over time and provides invaluable information to help develop predictive models; a necessity for successful prediction of water supplies from the high mountain headwater basins that supply much of western Canada and the northwestern US with rivers and life-giving freshwater.

ICE STATION PEYTO
Field drawing, pastel on paper, 18x24 cm, 2019.

Global Water Futures Observatories high elevation weather stations are also located on glaciers. This one on Peyto Glacier measures solar radiation, wind speed, air temperature and humidity. The tall metal pole drilled deep into the ice holds an ultrasonic radar on its arm to monitor the rate at which the surface elevation decreases from snowmelt and icemelt. These poles are drilled five metres into the ice and recently have had to be redrilled twice each summer as icemelt has increased to over seven metres per year.

PEYTO GLACIER
Oil on canvas, 18x24cm, 2019.

Scientific instruments are not needed to show regular visitors to the glacier that it is melting very rapidly. Its recession and reduction in volume is obvious to the human eye. This early summer painting shows deep meltwater channels, and the reddish tinge of deep purple algae which are feeding on deposits of ash from upwind wildfires. Canada had a record wildfire year in 2023.

FORTRESS MOUNTAIN RESEARCH BASIN HANGING TREE
Field drawing, pastel on paper, 30x24cm, 2022.

This well instrumented forest research station with a weighed, suspended tree - the hanging tree - is Tower Ridge Station, Fortress Mountain Research Basin. It is a cut, suspended, weighed fir tree to measure interception of snowfall and the storage and sublimation (snow transformed to water vapour) of intercepted snow in the subalpine canopy. This helps predict the impacts of forest cover change on water supply.

TRAIL VALLEY CREEK STATION
Field drawing, pastel on paper, 18x24 cm, 2019.

Trail Valley Creek Research Station (located in the Northwest Territories, Canada) was established in the early 1990s by Environment Canada and is now part of Global Water Futures Observatories. Currently led by Wilfrid Laurier University, it has grown to be the most instrumented and well-studied research basin in the Canadian Western Arctic, with several organizations conducting research there. Because of its remote location, the large range of equipment, and ongoing research projects, the station requires more support than most other observation facilities. These include accommodation and sleeping tents, field laboratories, and power supplied via wind turbines and solar cells. This station has documented dramatic climate change impacts on the snow hydrology and permafrost of this region over the last 30+ years.

2.2 Adaptation to the Changing Cryosphere and its Impact on Water Resources and Sustainability

Global heating is destroying all components of the cryosphere. The heating is amplified at high latitudes, and in most of these areas, we are observing reduced snow accumulation and snow cover duration, accelerated glacier mass loss and retreat, increased permafrost thaw, and advances in the timing and sometimes the rate of snow and ice melt. The areas experiencing these changes - circumpolar cold regions and high mountains - play crucial roles as major sources of global freshwater flow. Here, snow and ice play an important role in the local hydrology. Low-lying cold regions receive much of their precipitation as winter snowfall. Mountains receive greater amounts of precipitation than do lowlands and so are responsible for generating large amounts of runoff and streamflow.

The relative contributions of different cryosphere components to freshwater supply varies around the world. In most cold regions, the seasonal snowpack, rather than glaciers, is the primary source of runoff. Because the local characteristics and distribution of different cryosphere components have a strong influence on freshwater dynamics, the impacts of cryosphere destruction will not be uniform. The severity of the impacts for downstream communities will also depend on the relationship the communities have built with their water resources, and the therefore the consequences of changes in the volume, timing, and reliability of streamflow will depend upon water-use pattens. For agricultural producing areas, the loss of synchronization between the timing of mountain runoff and downstream demand for irrigation is a great concern.

Although not as significant as snowpacks to freshwater supply, glaciers offer tremendous value in terms of freshwater resiliency during periods of water stress. Glaciers melt fastest during the warmest, driest periods, and so especially once the mountain snowpack has been depleted, their rapid melt can partially compensate for otherwise reduced freshwater availability, helping to maintain streamflow until the end of the dry season. In regions where the dry season coincides with the growing season, glacier melt "drought-buffering" can be crucial for sustaining agricultural production. Many high mountain communities depend on water derived from glacial melt for food production, and glacial recession may force changes in these communities' historical practices or dependency on water stored in reservoirs. In some circumstances, recession may force communities to relocate altogether. As mountain glaciers recede and disappear, high mountain regions will lose the valuable ecosystem service they provide, decreasing resiliency against water stress.

Hydrometeorological station on Athabasca Glacier, Rocky Mountains, Alberta, Canada, and part of the Global Water Futures Observatories.

The impacts of cryospheric decline on water resources are complex and will vary across headwater basins and downstream regions. Declining snow accumulation and earlier snowmelt will cause earlier and less spring streamflow, followed by lower late season streamflow and greater dependency on rainfall and groundwater for water supplies. This will reduce the reliability of water resources and increase summer drought for vast downstream regions. Faster and greater glacier melt with climate heating can sometimes appear to compensate for diminished water supply from snowmelt in mountain streams, especially in the hottest driest periods, but is only a short-term supply of extra water that will diminish greatly over this century. The relative contribution of glaciers to freshwater supply also decreases with distance downstream. Communities living closest to glaciers are the most vulnerable to the impacts of glacier recession, though distant communities may still suffer from the loss of glacier drought-buffering services. Remaining snowpacks, groundwater and the impacts of thawing permafrost on river low flows are expected to grow increasingly important to water users as receding glaciers disappear.

ATHABASCA GLACIER RUNOFF
Oil on canvas, 91x116 cm, 2019.

The Athabasca Glacier, located in Jasper National Park, Alberta, Canada, is part of the Canadian Rocky Mountain Parks UNESCO World Heritage Site. It is fed by the Columbia Icefield, which also supplies ice to several other glaciers. The ice and snowmelt from the Columbia Icefield feed rivers which flow to the Pacific Ocean (Columbia River), Atlantic Ocean (Nelson River) and the Arctic Ocean (Mackenzie River).

UNESCO World Heritage Sites contain 10% of the world's glaciers and one-third of them will have disappeared by 2050. Ivanov paints a massive (see the tiny figures in the distance) meltwater runnel, and deposits of ash, soot and algae. The increasingly abundant ash and soot deposits from wildfires which deposit on the icefield and glaciers darken the surface and accelerate melt by up to 10%. Global heating has caused the Athabasca Glacier to recede 1.5 km and lose over half its volume in the last century. Jasper National Park was closed in 2024 because of catastrophic wildfires which burned down part of the Town of Jasper.

"This painting shows how menacing wildfires can be. It foreshadowed the truly shocking wildfires which afflicted many countries, including Canada, in the years following my putting oil on canvas."
- Ivanov

FIRE- GLOBAL WARMING'S SHOCKWAVE
Oil on canvas, 90x90 cm, 2019.

Ivanov's evocative painting reflects the more frequent, more intense and larger wildfires burning forests, grasslands, tundra, homes, and communities caused by climate change. The Transitions team saw widespread evidence of forest burns and the consequent large depositions of ash and soot on glacier surfaces, sometime hundreds of kilometers from the fire.

Wildfires are an important part of natural ecosystem renewal and small wildfires were part of how Indigenous peoples managed forests and grasslands and prevented larger wildfires. Modern global heating-driven wildfires are large, intense and dangerous and can also cause increased flooding after fires have burned soils and vegetation and so reduce the natural ability of river basins to retain and evaporate water from snowmelt and rainfall.

COLD REGIONS WARMING
Oil on canvas, 100x150 cm, 2020.

The vast cold regions of the circumpolar north extend far southwards of the Arctic Circle. The Northern Hemisphere winters and summers are getting hotter. The Arctic sea ice is thinning and contracting. The Greenland icesheet is melting at unprecedented rates. Vast tracts of the permafrost (permanently frozen ground) are thawing. Vegetation and peat fires are now extensive. The red hues in this painting of the circumpolar regions are an artistic metaphor for these dramatic changes. The reduction in ice cover reduces the reflectivity of high latitudes, leading to greater absorption of energy from the Sun, leading to greater heating; a powerful positive feedback loop.

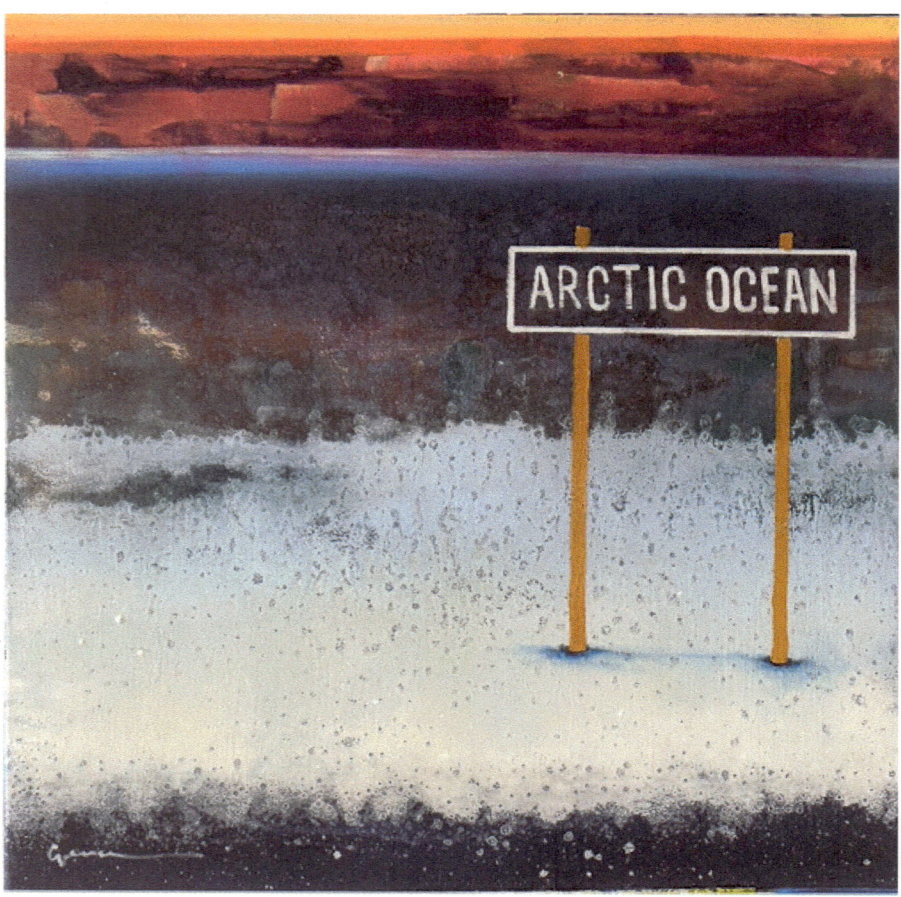

FIRE ACROSS THE ICE
Oil on canvas, 80x80 cm, 2020.

The Arctic Ocean has undergone a massive warming in the last 40 years, with the loss of over half of its multi-year sea ice. This ocean is surrounded by the Arctic lands of Russia, Canada, Greenland and Scandinavia – much of which are warming three to five times faster than the rest of the world – and receives the warmer and increasing freshwater flows from the major northward flowing rivers of Russia and Canada such as the Ob, Yenisei, Mackenzie, and Lena as well as freshwater from melting glaciers and ice sheets. Ivanov's view from the sign at Tuktoyaktuk, Northwest Territories, Canada, looks across the remaining ice of the Beaufort Sea, foretelling the increasing temperatures, permafrost thaw, greenhouse gas emissions, wildfires, floods and streamflow volumes that span the circumpolar North. The bubbles in the ice remind us of methane emissions from permafrost through lake ice that are ubiquitous in northern peatlands.

LONELINESS
Oil on canvas, 101x101 cm, 2023.

The Great Thaw is impacting many species. To many people, it is the polar bear which epitomizes the threat: sea ice is declining in much of its range, seriously affecting its ability to hunt.

POWERFUL AND VULNERABLE - THE LAST BEAR
Oil on canvas, 100x100 cm, 2018.

This reflects the last stand of nature as the melting cryosphere leaves many species stranded. The polar bear is an iconic symbol of a cold regions animal that is threatened by declining sea ice in much of its range. It needs the sea ice to hunt and so looks vulnerable without it. This is an almost inevitable outcome in the absence of rapid and deep cuts in global warming gas emissions.

THE RIVER IN THE SKY
Oil on canvas, 130x170 cm, 2021.

This painting reflects the record-setting rainfall, extensive flooding and destruction of transportation infrastructure, homes and farms that occurred because of an "atmospheric river" of heavy rainfall falling onto melting mountain snowpacks in November 2021 in southwestern British Columbia, Canada. The resulting floods were the costliest in BC's history at $US 7.5 billion and were devastating to farms in the Fraser Valley, where more than 600,000 farm animals perished. Global heating means that the atmosphere can carry more moisture and so extreme precipitation events are becoming even more extreme.

THE DENIER AND THE BEAR
Mixed media on hessian, 175x135 cm, 2019.

Despite overwhelming scientific evidence, there are still people who deny climate change and/or the role of humans. The consistent evidence comes from many sources, including Global Water Futures Observatories monitoring stations, such as the one represented here, at Trail Valley Creek Research Basin in Northwest Territories, Canada which has support and accommodation tents. The Polar Bear, which – to many – is symbolic of the threat which climate changes poses for many animals, has supplemented his already impressive weaponry. He needs all the help he can get.

Global Water Futures Observatories

The Global Water Futures Observatories (GWFO) is Canada's premier freshwater observation network, operating 64 instrumented monitoring sites across the country's major river basins. These observatories generate open-access, high-resolution data on climate, glaciers, hydrology, and water quality—essential for monitoring changes, diagnosing risks to water security, and guiding evidence-based policymaking. With climate change accelerating water crises, GWFO's data directly informs disaster preparedness, agricultural resilience, and sustainable water management, equipping governments with the tools to safeguard communities from floods, droughts, and water contamination. With growing pressures on water resources, this national infrastructure plays a crucial role in supporting Canada's economy, public health, and environmental resilience.

Learn more at **www.gwfo.ca**

© John Pomeroy

3. Mountains and Forests

Snow-covered forest, Canadian Rockies.

3.1 Mountains

Mountain cold regions form the headwaters of many rivers around the world and so play a major role in the global hydrological cycle. Cyclical warm-season melting of mountain snowpacks and glaciers releases freshwater, which can flow directly into streams and rivers or can percolate into the ground, replenishing soil moisture and groundwater.

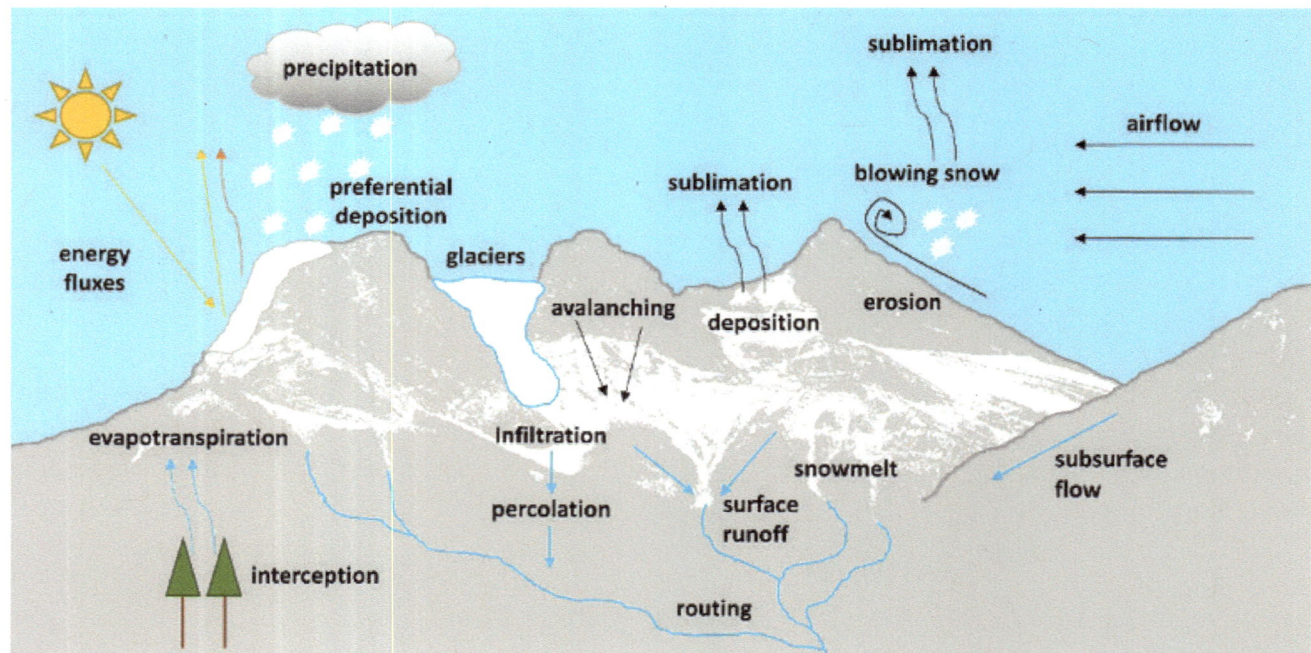

Figure 3.1: Mountain hydrological and cryospheric processes governing water supply. Adapted from © Bertoncini (2024) with permission.

Glaciers

Glaciers form from the accumulation of perennial snowpacks that, over time, compress into ice. A glacier's "mass balance" is the net difference between snow accumulation and snow and ice melt; a glacier with a positive mass balance will grow, whilst one with a negative mass balance will retreat. Almost all glaciers around the world have recorded negative mass balances. Glacier melt and sublimation are strongly impacted by snow cover duration, solar radiation and temperature. In turn, these are affected by air temperature, cloud cover, ice surface reflectance (also called 'albedo'), winter snowfall, and snow redistribution by wind and avalanches and so are very sensitive to climate change.

Glacier melt is not only affected by climatic parameters. Dust, soot, and microbial and algal growth on snow and glacier surfaces can accelerate melt rates by decreasing surface albedo. The effects of debris cover are complicated, though, because if a debris cover is more than a few centimetres thick, the underlying ice mass can become insulated against heating and end up persisting even after the glacier appears to have receded. "Dead ice," as it is called, contributes to the complexity of high mountain hydrology. The presence - or absence - of glaciers affects other systems, too, as glaciers can enhance snow accumulation, delay streamflow generation, and reduce stream temperatures when they melt. Figure 3.2 summarizes some of the expected shifts in high mountain environments due to atmospheric warming.

Hydrometeorological station on Athabasca Glacier, Rocky Mountains, Alberta, Canada, and part of the Global Water Futures Observatories.

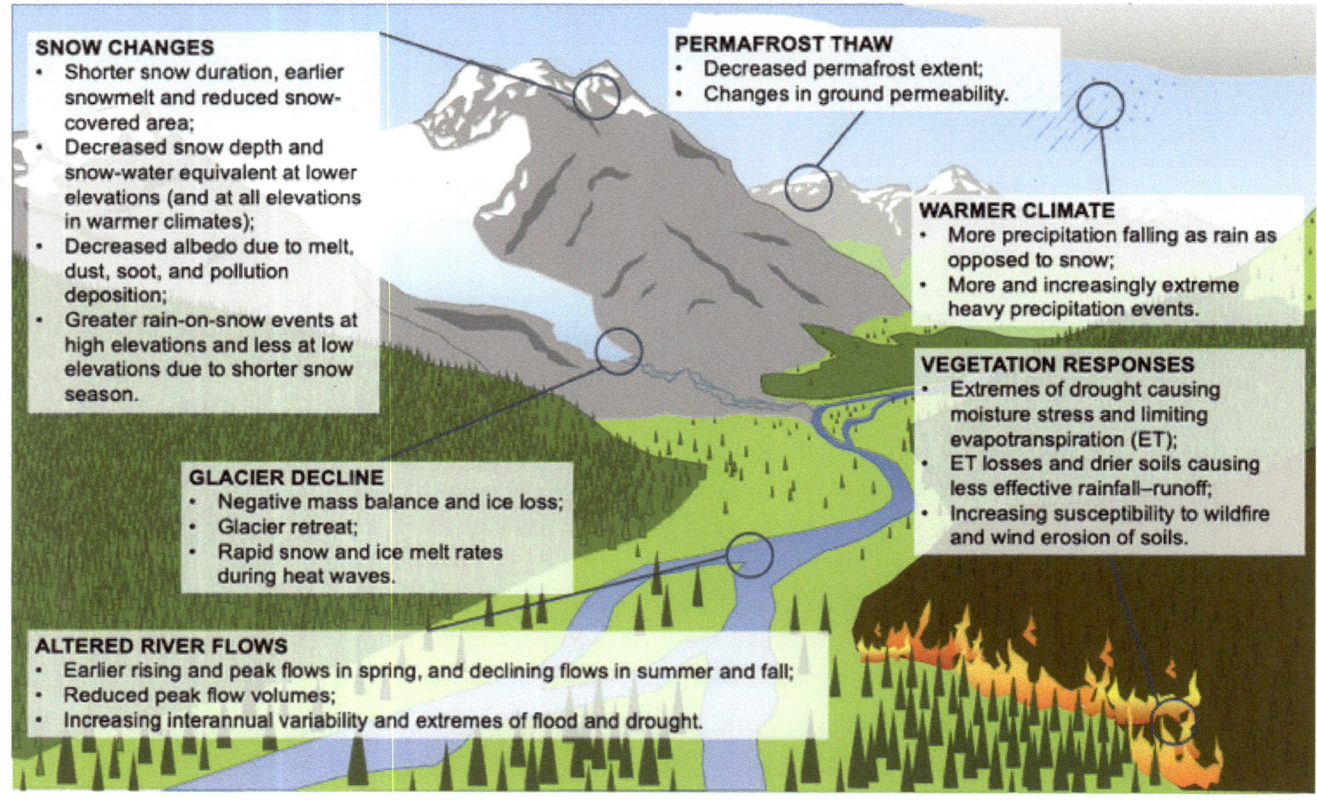

Figure 3.2: Conceptual illustration and summary of key cryospheric and hydrological changes in mountain and forested regions under climate warming of the 21st century. © Chris DeBeer, GWF, University of Saskatchewan.

MOUNT LOGAN
Oil on canvas, 100x100 cm, 2021.

A studio painting which shows spectacular snow-covered and still-intact ice and glaciers of a mountain rising 5,959 metres above sea level; a scene which is at risk at lower elevations around the world.

STILL FROZEN, FOR NOW
Field drawing, pastel on paper, 30x42 cm, 2019.

The Saint Elias Icefield is the third-largest icefield in the world, after Antarctica and Greenland. The ancient ice-sculpted forms of distant mountains include Mount Logan. They are a spectacular subject being pointed out by the distant figure of Professor Pomeroy. Despite having the greatest elevational range in the world, the Saint Elias mountain range has lost a quarter of its ice cover in the last 50 years. The icefield still feeds the receding Kaskawulsh Glacier that used to support the now diminished Kluane Lake, which is shown later.

PEYTO GLACIER
Oil on canvas, 18x24 cm, 2019.

Peyto Glacier in the Canadian Rockies of Alberta, Canada, is one of the world's longest-studied glaciers. It has lost more than 70% of its volume since the beginning of the 20th Century with the most rapid loss being in the last decades. It is losing 3.5 million cubic metres of water each year. Observation stations placed on the glacier in recent years have been lost because the ice is melting so rapidly. Where there was once ice, there are now banks of silt and mud and a new proglacial lake, named "Lake Munro" by scientists from the University of Saskatchewan, currently studying the glacier, in honour of retired Peyto glaciologist Professor Scott Munro.

UNDER THE TOE
Oil on canvas, 18x24 cm, 2021.

Peyto Creek, meandering through the recent glacial deposits, under a smoky sky. The heat dome over western Canada in July 2021, produced the highest ever temperature then recorded at Peyto: 23 °C. Peyto Glacier retreated over 190 m and melted 7 m downward in 2021 as a result. The prolonged hot, dry spell also triggered and maintained many forest fires.

CHANGING PALETTE ON PEYTO
Field drawing, pastel on paper, 30x42 cm, 2019.

This view down-valley from a former position of the glacier snout shows the Government of Canada's research huts and weather stations, which were installed conveniently near the glacier's edge in 1965. They are now stranded atop the bluff on the left more than 1km from the ice and hundreds of metres above it.

LOST GLACIER AND NEW LAKE
Field drawing, pastel on paper, 33x45 cm, 2019.

This lake would have been under ice fifteen years ago. The artist is looking down from the present (2019) toe of Peyto Glacier at the newly-formed and rapidly expanding "Lake Munro" and Peyto Creek that flows down the U-shaped valley to Peyto Lake hundreds of metres below.

A RAPIDLY-CHANGING LANDSCAPE CAPTURED IN AN INSTANT
Field drawing, pastel on paper, 18x24 cm, 2019.

One of Ivanov's field drawings; a fifteen-minute production, witnessed by the scientists.

Mountains and Forests

PEYTO CREVASSES IN BRIEF
Field drawing, pastel on paper, 18x24 cm, 2019.

Two of Ivanov's rapid field drawings; part of the preparation for the Peyto Crevasses oil painting (see page 59).

Peyto Glacier forefield, the barren area exposed by glacial melt.

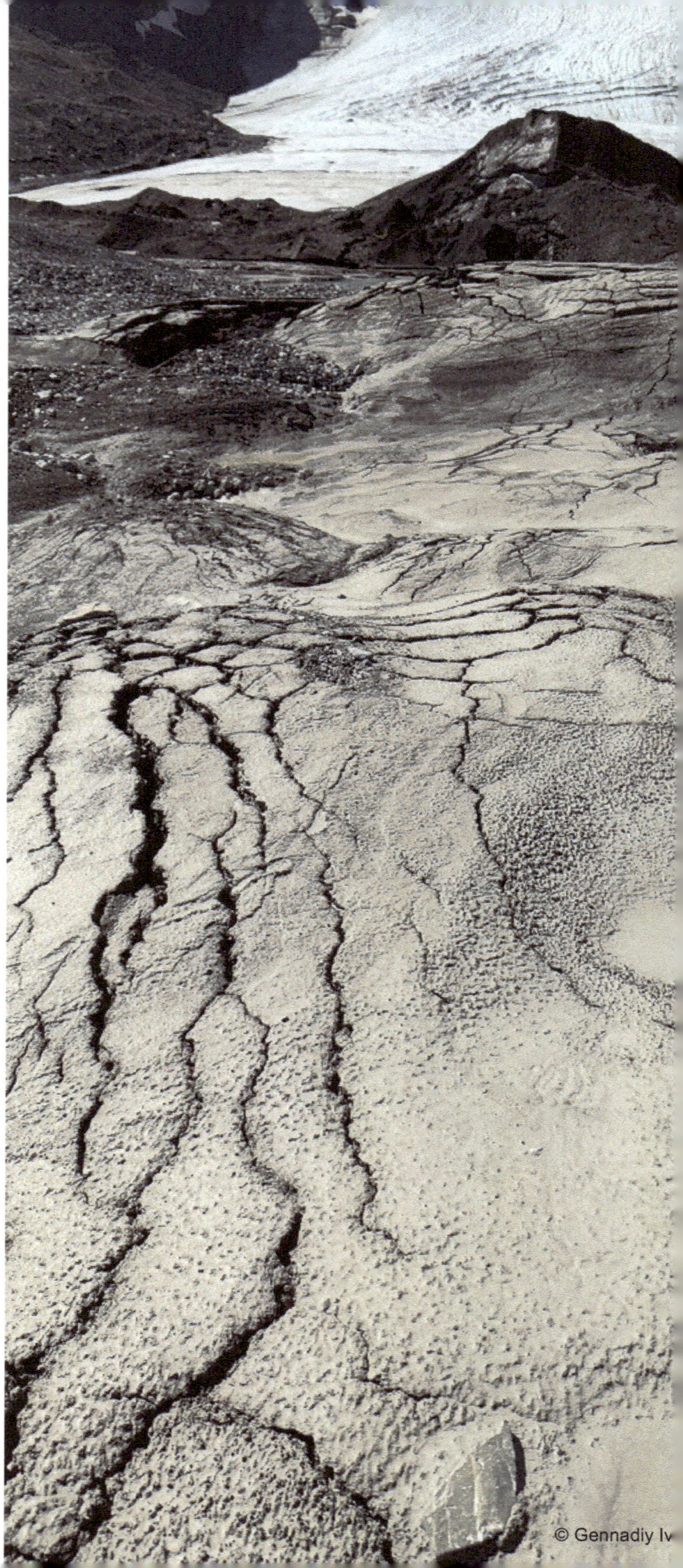

© Gennadiy Iv

Data recording at Peyto Glacier, Rocky Mountains, Alberta, Canada, and part of the Global Water Futures Observatories.

REQUIEM FOR PEYTO
Oil on canvas, 100x80 cm, 2021.

Mass balance records for Peyto Glacier go back to 1965 and records of its length to 1897. These and other data records are archived at Global Water Futures Observatories, the Geological Survey of Canada, the World Glacier Monitoring Service and the Global Cryosphere Watch. They are invaluable documentation of a rapidly shrinking glacier.

"I enjoy surreal painting. It helps me express my emotions; something which is important to me. I know that scientists also have emotional responses to what they are seeing and studying. But, in their public statements, they are careful to express themselves in objective terms, based on the rational methods and reporting of science. Because I am an artist, I am allowed to portray myself in a way which expresses some of my feelings.

Here I am below the current snout of the Peyto Glacier, amidst the new and barren landscape revealed by the glacier's rapid retreat. Although barren, the newly-emerged post-glacial depositional landscape does show tiny specks of green – the first plants are already moving in. They are shown in my glass. Also in my glass is the beige-yellow glacial silt of the depositional landscape, and cryoconite. Cryoconite, the scientists explained to me, is a cocktail of materials which accumulates each year on the glacier's surface. It consists of ash and soot from vegetation fires, algae, bacteria, viruses. It has been growing in abundance over the years, accelerating the glacier's decline, and is washed-off by the annual melt-water to form dark deposits below the snout. It aids the growth of seedlings and moss. At what point in the future will the blue-white icescape behind me be transformed to green?

I also record the sound of the glacier. The ice-driven katabatic wind; the wind-driven snow particles in late winter; the torrents of meltwater in summer; the splitting and crashing of the collapsing glacier. The record-player is my surreal expression of this. It is also a way for me to emphasize the importance of the painstaking recording of scientific data. Peyto observations started more than 120 years ago, making it the longest-studied glacier in North America, In another 120 years there will only be the record left." - Ivanov

PEYTO CREVASSES
Oil on canvas, 91x116 cm, 2019.

Peyto Glacier has developed extensive crevasse networks in its lower toe area as the ice stagnates and melts at the fastest rate ever. These crevasses conduct meltwater to the bottom of the glacier where they rush out the snout to form Peyto Creek that flows into Peyto Lake, then to the Mistaya River and eventually the North Saskatchewan River. The North Saskatchewan River waters contribute to the Nelson River which flows to Hudson Bay.

ICE FLOW
Oil on canvas, 13x18 cm, 2019.

TO THE SASKATCHEWAN RIVER DELTA AND THEN HUDSON BAY
Oil on canvas, 91x116 cm, 2019.

Glacier ice flows from the accumulation zone where the snow cover does not fully melt, to the ablation zone where much older ice is exposed and melts. Ivanov has used heavy applications of oil paint to evoke the flowing ice of Peyto Glacier. Although it is decaying, and facing death, there is still a sign of life.

Seen from a position almost on the toe of the glacier, Peyto's meltwater eventually contributes to the flow of the North Saskatchewan River, an important source of water for the Saskatchewan River Delta, the largest freshwater delta in North America and the home of a critically threatened wetland ecosystem and an Indigenous community, Cumberland House, that has found itself in a water crisis due to lack of supply in 2013. The demise of such mountain glaciers will have far-reaching impacts downstream.

A SIGN OF DEATH

Pastel on paper, 45x65 cm, 2019.
Oil on canvas, 160x160 cm, 2021.

Ivanov saw a skull in the crevasses and broken ice masses near the toe of the glacier, tinted by the remains of algae, soot and ash. He took this as a portent of the glacier's future: no future.

PEYTO'S WOUNDS
Oil on canvas, diptych, both paintings, 160x160 cm, 2021.

"To produce this diptych, I have relied on field drawings and photographs, my memory, and very high-resolution stills and videos from drones operated by GWF scientists on that day. The meltwater was flowing in runnels and torrents across and through the ice, and parts of the glacier were dramatically coloured, especially by the cryoconite (accumulations of ash and soot from wildfires, fungi, bacteria, as well as algae). At places at the margins of the glacier, but especially below its receding snout, were deposits of yellow-tinged glacial silt. The whole scene taken together, including the bare moraines and sediment beyond the snout, summoned - in my mind - a sense of destruction.

"The red hues signal scars which, I imagine, represent the bleeding and screaming of this ancient, moving, and ever-changing entity. Despite the vivid colouration, it represents a step closer to final death, decay, and darkness. I visualize the yellow tinges as the transformation of progressively deeper, bleeding scars to the pervasive dirty yellows which will be the remaining ice-free depositional landscape and which, for a short time, will be desert-like in its absence of vegetation. Peyto is doomed, a fate perhaps camouflaged by this transient brightness." - Ivanov

STARING AT COLLAPSE
Field drawing, pastel on paper, 18x24 cm, 2019.

Meltwater rushing through a recently exposed sub-glacier meltwater channel seen from atop a lateral moraine gives a foresight of things to come. Within two years the channel cavity had collapsed, causing a local drop in the elevation of the glacier surface of 56 m which far exceeded that caused by melting in other parts of the glacier. Above left is a map of glacier surface elevation changes from 2019-2024, precisely measured by LiDAR (Light Detection and Ranging) from a drone. The map indicates melt rates of 30 metres and ice lower rates of up to 56 metres over five years.

Above: Drone with LiDAR sensor.

Right: Change in Peyto Glacier surface elevation from 2019 to 2024 as measured by LiDAR in late summer. © Madison Harasyn, Centre for Hydrology, University of Saskatchewan

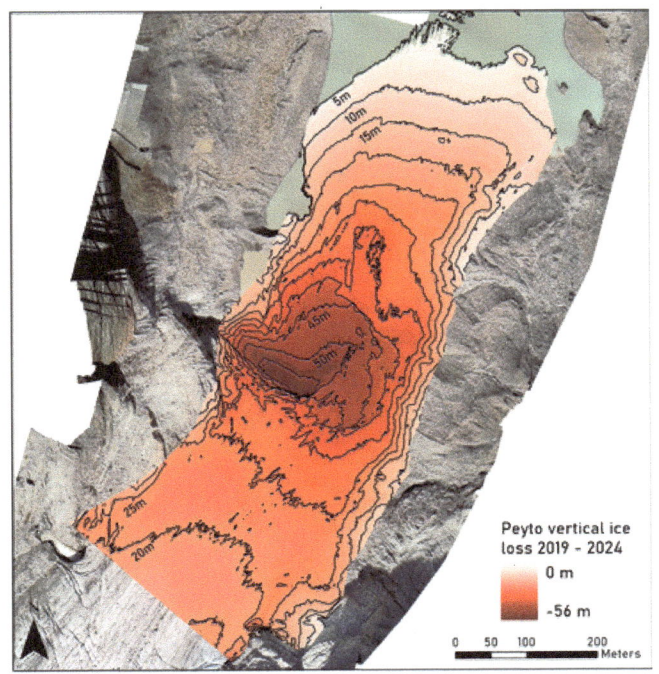

BIRTH OF A RIVER, DEATH OF A GLACIER
Oil on canvas, 91x116 cm, 2019.

The rapidly retreating Peyto Glacier reveals a former sub-glacial channel that is emerging as a large stream beside the glacier. The stream is fed from melting ice and snow and is choked with ice that has collapsed from tunnel walls within the glacier and now will flow with the river to melt downstream. The cold water from these streams provides ideal conditions for native trout in the Canadian Rockies. This cold water has become even more important as recent hot summers have warmed river temperatures above the cool conditions that trout require. Glacial meltwater is an important contribution to river flows in years of drought.

Part of Peyto Glacier collapsed, exposing a former sub-glacial channel. August 2019.

PEYTO IN FIVE YEARS
Oil on linen canvas, 121x182 cm, 2024.

The melt is accelerating, with the impression that the water is being sucked into a black hole, and soon there will be nothing left. The black hole is redolent of the watery ice caverns - caused by rapid melting and ice collapse - in the previous paintings.

GLACIER DECLINE
Charcoal on paper, 180x100 cm, 2019.

Fore fronting the now-exposed strata which were once the side of the glacier valley are deposits of glacial silt which have accumulated below the glacier snout. The silt deposits are cracking in the summer heat.

Pomeroy likes this drawing because "It gives the illusion of the glacier transforming into a river delta – it speaks to glacial hydrology and the loss of these glaciers and their replacement by terrestrial hydrological systems. And that sediment and cracking can also be dangerous….", referring to a many fatalities in Uttarakhand, India, caused by the collapse of part of a Himalayan glacier and rock well above the river in 2021.

GLACIER DECLINE; EMERGENCE OF STRANGE NEW LANDSCAPE
Oil on canvas, 150x100 cm, 2020.

Peyto Glacier in Alberta, Canada, is one of the world's longest-studied glaciers. It is rapidly receding because of climate change, having lost more than 70% of its volume since the beginning of the 20th century and has retreated almost half of a kilometre since 2019. Here, the remaining ice is seen in the distance, behind strange deposits of black cryoconite, and glacial silt in the foreground. The cryoconite accumulated on the ice surface and is then washed off by the copious summer meltwater; it consists of ash and soot from wildfires, bacteria, fungi and algae. Each year it darkens the glacier's surface, reducing its reflectivity, and exacerbating melt. The deposits of glacial silt are cracking in the summer heat.

WILD BILL WOULDN'T RECOGNISE IT
Oil on canvas, 90x150 cm, 2019.

The Peyto Glacier is named after "Wild" Bill Peyto, who was born in England. Upon moving to Canada, he became, from the 1890s onward, a: pioneer, railway labourer, trapper, prospector, horse outfitter, packer, legendary mountain guide, and eventually one of the first wardens of Banff National Park. Bill Peyto would have seen a glacier almost 2.5 km longer than the current remnant ice mass. Where there was once ice, there are now banks of silt, mud and cryoconite. In the distance, perched on a bank of mud, is a form of transport Bill would not have recognized either.

THE BLACK MOUNTAINS AT THE END OF A GLACIER

Field drawing, pastel on paper, 18x24 cm, 2019.

The accumulations of cryoconite, washed off the surface of the glacier, form black piles that look like miniature mountain ranges, 1-2 metres high. The depositional landscapes below the glacier's toe are strange black mountains, new to the Earth.

ICE, WATER, ROCKS, CLAY, SILT AND BLACKNESS
Oil on canvas, 80x80 cm, 2019.

Ivanov captures the strangeness of the deglaciating environment. The retreating glacier leaves ponded meltwater, rocks that have been eroded and transported from the mountainsides, clay deposits from fine materials ground by the glacier and left as sediment in lakes, silt from ground rock left in glacial meltwater channels and blackness of a future without snow and ice - one of extreme heat and hostility to life.

CRYCONITE SEDIMENT I, II
Oil on canvas, 80x60 cm, 2019.

"One of the most fascinating outcomes of conversations with scientists has been my growing realisation of how interconnected the world is and how something that is very small can affect the whole planet." - Ivanov

WORLDS WITHIN WORLDS
Oil on round canvas, 20 cm diameter, 2019.

There is a strong connection between the rate of melting of the ice and the "brightness" (albedo, in scientific terms) of the ice surface. Clean ice is very bright and melts slowly, whereas dark ice absorbs more solar energy and melts more quickly. Rapid ice melt leads to sea level rise and deglaciation. This dark material is known as cryoconite, and is teeming with life, including pollen and living organisms such as algae and bacteria. Cryoconite accelerates glacier melt. It can be studied through the technique of scanning electron microscopy (SEM), which produces images at, typically, around 10,000 magnification. These paintings are based on SEM images of cryoconite samples from the Peyto Glacier – collected and analysed by Global Water Futures scientists at the University of Saskatchewan Toxicology Centre. This normally unseen "microworld" has profound impacts on our Earth - the accelerated melt caused by these micro-organisms contributes to global sea level rise.

A DELUGE OF LAVA AND ICE – THE END OF PEYTO GLACIER
Oil on linen canvas, 150x100 cm, 2021.

A blood red, lava-like flow replaces the glacier and its meltwater, representing the rapid, catastrophic melt in the record the record hot summer of 2021 when Peyto Glacier retreated 200 m, roughly ten times its recent rate. Whitehorse, Yukon, experienced record flooding from rapid glacier melt upstream. More than 600 people died in British Columbia from the record heat, where temperatures reached 49.6 °C in Lytton just before the town burned to the ground in a wildfire.

MOUNT ATHABASCA

Oil on canvas, 30x30 cm, 2019.

The glaciers and snowpacks of Mount Athabasca are not what they used to be and the routes to climb this mountain that were viable in the 1980s are no longer suitable. It is a place of snow-covered crevasses and avalanches and remarkable beauty. Here, remnant snowpacks and glacier ice remain before the melt cascades of the spring freshet. What will be left in another 40 years?

ATHABASCA WARMING

Oil on canvas, 30x30 cm, 2019.

Here the dark red sky portends the increasing global temperature over decades which is driving many glaciers to extinction. The sky also indicates the smokiness from distant forest fires, deposits from which increase the absorption of solar radiation at the glacier surface. The glacier surface shows cryoconite – formed from soot, dust and algae and further darkening the surface – and meltwater rushing off the ice. This glacier melted downward at 4 to 5 m per year about 10 years ago, but recent record high temperatures have increased this downward melt rate to 6 to 8 m per year. Even with its current thickness of the height of the Eiffel Tower (330 m), it can only last a few more decades under such rapid melt.

ATHABASCA HEATING
Oil on canvas, 20x20 cm, 2019.

The Athabasca Glacier normally experiences an Icefield Wind - a windy, cold, 100 m thick layer of air draining from the Columbia Icefield heights down towards its toe. This cool wind has helped to slow the melt of the Athabasca Glacier. But recently, warm air is penetrating this stable air layer and temperatures exceeding 15 °C are now common on the ice. The red sky painted here shows this warming and rapid melt – dangerous days ahead for this iconic glacier.

ATHABASCA GLACIER
Pastel on paper, 38x58 cm, 2019.

In the lower middle of Ivanov's field painting are moraines which reveal the extent of glacier recession. The ice itself has a pink hue associated with algae. It has retreated by about 600 m since the 1960s and is now thinning at around 7 m per year. Whereas Peyto Glacier is regarded as the bell-weather glacier for scientists because it has been continuously observed since the 1960s, with a good photographic record back to the 1890s, it is the Athabasca Glacier which is the popular canary-indicator since it is so easily accessible to tourists. A bus even transports visitors onto the glacier. It is also subject to intensive scientific study by many universities and monitoring by Parks Canada.

SUMMER AT MOUNT ATHABASCA
Pastel on paper, 38x58 cm, 2019.

Summer in the Canadian Rocky Mountains UNESCO World Heritage Site is normally a time of vigorous vegetation growth and evapotranspiration as shown by the green trees and wildflowers in front of Mount Athabasca. Water for vegetation growth, groundwater recharge and streamflow generation is provided by summer rainfall and the melt of snowpacks and glacier ice. In recent droughts caused by global heating, snowmelt came too early, causing a longer summer in which rainfall could not support vegetation growth and streamflow adequately. The results were record low streamflow and soil moisture and record high wildfires and glacier ice melt. Together these are destroying the character of this World Heritage Site – a loss for humanity and for the planet. Similar impacts are being felt in the 50 World Heritage Sites around the world that contain 18,600 glaciers. By 2050 glaciers in one-third of Word Heritage Site glaciers will have disappeared including the last glaciers in Africa, the Pyrenees and Yosemite Park, California, USA.

Athabasca Glacier in 1980 (top) and 1986 (bottom); these scenes have disappeared along with much of the ice since these photographs were taken on John Pomeroy's early visits to the glacier.

UPPER ATHABASCA GLACIER AND THE COLUMBIA ICEFIELD
Field drawing, pastel on paper, 33x45 cm, 2019.

ATHABASCA MELTWATER CHANNEL
Field drawing, pastel on paper, 33x45 cm, 2019.

Here Ivanov has produced a field pastel of the icefall from the Columbia Icefield that feeds the Athabasca Glacier. The contrast in hues from the colder, cleaner blue ice and perennial snowfields of the Icefield to the Athabasca Glacier whose surface has been darkened by wildfire soot, dust, and algae is clearly apparent. Surface darkening of the glacier increases its melt rates by up to 10% in years with severe wildfires. And wildfire occurrence is increasing as is their severity – a further consequence of global heating.

The deep meltwater channel on the glacier surface inspired Ivanov to produce several oil paintings. Meltwater from the Athabasca Glacier feeds the Sunwapta River and eventually the Athabasca River which contributes to the Mackenzie River on its journey to the Arctic Ocean near Inuvik, Northwest Territories, Canada. Streamflow on the Sunwapta River is dominated by glacier melt and so increases in the hottest, driest part of summer – the inverse of non-glacial fed rivers in the region and an important "drought-proofing" of mountain streams. Unfortunately, this has been insufficient to prevent the Athabasca and Mackenzie Rivers from sustaining record low summer flows in recent years. The flows have been so low that barges cannot travel the Mackenzie River to resupply the communities of the Western Arctic – a disastrous consequence of hydrological drought associated with global heating that even enhanced glacial melt cannot compensate for.

DEEP MOULIN ON THE ATHABASCA GLACIER

Oil on canvas, 108x150 cm, 2019.

Large moulins (or mill holes) which deliver surface meltwater to the base of the glacier are common sights in the summer. They are also exceedingly dangerous to those traversing the ice. During the "Heat Dome" and exceptionally hot summer of 2021, the Athabasca Glacier melted faster than ever before recorded, and peak streamflow discharge on the Sunwapta River was 80% higher than the long-term average. Summer discharge generated by the Athabasca Glacier was 36% higher than the long-term average. This is an example of the peak water regime from a rapidly receding glacier. The downstream drought proofing from this rapid glacier melt in times of extreme heat will not be available once the glacier has melted further and receded to a small ice mass in a few decades.

FOREFIELD OBSERVATION STATION I, ATHABASCA GLACIER
Field drawing, pastel on paper, 45x33 cm, 2022.

Ivanov noticed the astonishing deterioration of the glacier from his first visit in 2019. The three lakes below the toe of the glacier have grown considerably. Scientists rely on a range of observation techniques to quantify the changes, and to develop prediction models The observation techniques include automatic weather stations such as this one from Global Water Futures Observatories, but also satellite images and airborne lidar. The figures of Professors Davies and Pomeroy can be seen in this pastel, discussing the rapidly changing environment.

FOREFIELD OBSERVATION STATION II, ATHABASCA GLACIER
Pastel on paper, 47x62 cm, 2019.

ATHABASCA ICE STATION
Pastel on paper, 47x62 cm, 2019.

This station was set up by the University of Saskatchewan Centre for Hydrology in 2013 to provide a baseline of the climate downwind of the Athabasca Glacier and Columbia Icefield. It now forms part of the pan-Canadian Global Water Futures Observatories network of instrumented basins – a contributor to the World Climate Research Programme. Besides temperature, wind speed, humidity, snow depth and precipitation, it also measures soil temperature and moisture content. The location was glaciated earlier in the 20th century. Such measurements are vital to understand the newly developing postglacial climate.

The University of Saskatchewan Centre for Hydrology installed a hydrometeorological station on the ice of Athabasca Glacier 2003. This is several hundred metres above the on-glacier tourist bus "turn-around". Measurements of air temperature, humidity, wind speed, radiation, and snow depth are relatively straightforward; but recording ice surface height is very challenging. The ice surface is melting downwards at about 7m per year. The algae, and the soot and ash deposits from wildfires, are accelerating the melt rates along with record warm summers and shorter winters.

FARE THEE WELL, ATHABASCA?
Oil on canvas, 18x24 cm, 2019.

Ivanov has moderated the sky colours in this oil painting of the glacier. It is his plea for humanity to restrain global heating and to preserve glaciers through a responsible world view of the importance and role of mountain glaciers and snowpacks. About 10% of the planet's surface is covered by ice masses and glaciers store about 75% of the world's freshwater.

ANGELS & DEVILS I, II
Oil on canvas, 90x60 cm, 90x70 cm, 2020.

The Angel Glacier of Jasper National Park is iconic and a major feature of Mount Edith Cavell, itself named after a heroic British nurse of the First World War who was viewed by many as an "angel" in a dark time. The glacier is so-named because, from a distance, it appears to have the outspread wings of an angel. It is beset by burning forests and wildfire in these impressionistic paintings showing that both fire and ice are important features of this era of rapid global heating. The icefall of part of the adjacent Ghost Glacier into Cavell Lake in 2012 also represents the darker side of glaciers. The icefall triggered a small tsunami that destroyed the parking lot and access road and trails in this popular area. Fortunately, it happened at night with no loss of life.

STANLEY GLACIER FIELD PASTEL

Field drawing, pastel on paper, 33x45 cm, 2019.

Stanley Glacier is in Kootenay National Park in British Columbia, in the headwaters of the Columbia River which supplies water for ecosystems, food and energy in vast areas of British Columbia and the US Pacific Northwest. Although not as intensively monitored as glaciers like Peyto or Athabasca it, too, is receding rapidly. The glacier is in a forested valley burned by the Vermilion Pass forest fire which destroyed 2,500 hectares over 18 days in 1968, and by subsequent fires in 2003 and 2018.

STANLEY GLACIER
Oil on canvas, 100x100 cm, 2021.

The Burgess Shale deposits in the valley overlooked by Stanley Glacier reveal 508-million-year-old fossils that show the strange and wonderful life after the "Cambrian Explosion" of biological diversification. Most of these organisms became extinct about 20 million years later, during a glaciation. The Burgess Shale was part of the reason that 2,360,000 ha of the Canadian Rocky Mountain Parks were declared a UNESCO World Heritage Site in 1980. The rapidly receding Stanley Glacier is in the headwaters of the Columbia River Basin which supports ecosystems, irrigation, hydroelectricity and water consumption of a vast area (668,000 km²) of British Columbia and the US Pacific Northwest.

THE GLACIER IS ROCK!
Oil on canvas, 90x150 cm, 2019.

One of the more remarkable sights in Yukon, Canada, are "rock glaciers" – debris covered ice masses. These are left when the glacier ice remains beneath the rocky debris. The melt rate of debris covered ice is far slower than that of exposed glacier ice, preserving these "rock glaciers" far longer than the ice glaciers that once fed them. The rocks were once carried within or on the glacier, and have been added to by fall from the valley sides. There are ripples in the surface, remnant from the flow of the former glacier or a result of continuing slow movement.

GHOST GLACIER
Oil on canvas, 91x116 cm, 2019.

Inspired by the Rock Glacier in Kluane National Park, Yukon Territory, Canada, Ivanov paints the ghost of a glacier in the distance - sitting above the ribbon of rock - shrouded in blowing snow which is suspended by the wind from one of the north-facing sides of a mountain gully. In the foreground are the waters of Dezadeash Lake, which is still drip-fed by the waters flowing beneath the rock glacier, the rocky remains of what was once a glacier of ice. Debris-covered ice may be all that remains of many mountain glaciers by the end of this century due to enhanced melt from global heating. They melt much more slowly than ice glaciers and so may remain for centuries after the disappearance of the exposed ice.

PRESERVED GLACIERS?
Oil on wooden panel, 60x80 cm, 2019.

Ivanov produced a series of evocative paintings of retreating glaciers inspired by the International Year for Glaciers' Preservation 2025. The need to stop this retreat and preserve the glaciers comes through in his artwork – a reset that can be accomplished by reducing the concentrations of greenhouse gases in the atmosphere by cutting humanity's emissions.

Snow

In most high latitude and altitude environments, the formation of the seasonal snow cover provides most freshwater storage. There are many processes governing the distribution of snow and the resulting melt patterns. In addition to climatic factors such as wind, humidity, and incoming radiation from the Sun, where and how snowpacks accumulate is influenced by topography, sun exposure, the presence and characteristics of vegetation, and more.

There are many paths for snow to take when it falls in mountain environments. Snow may be caught by forest canopies, where it might then evaporate directly back into the atmosphere, drop off the branches some days later, or melt, come warmer temperatures. Snow may be blown over vast distances by wind or avalanche off mountainsides, ending the season far from where it originally fell from the sky. Global heating is changing how snow accumulates on the ground through shorter snow seasons, less snowfall and more rainfall, wetter or icier snow that is less readily blown by the wind and by increasing vegetation in cold regions. For instance, warmer Arctic winters with expanding tundra shrubs can reduce blowing snowstorms and increase the depth of snowpacks, but warmer sub-arctic winters with expanding needleleaf forests, conversely, can intercept more snow in the thickening forest canopies and decrease the depth of snowpacks.

During melt, some snowmelt water may infiltrate into the ground, while other meltwater may runoff the land surface into streams and rivers. The path meltwater takes depends on how fast the snowpacks melt, whether there is underlying soil or rock, the steepness of mountain slopes, and more; understanding snow dynamics is not simple. Snowmelt is slowed by increasing vegetation and melts two to three times more slowly under a dense forest canopy than out in the open. Snowmelt is especially sensitive to climate change, which can increase or decrease precipitation and increase both air temperature and humidity, causing more frequent winter rainfalls, advancing the timing of snowmelt, causing changes in rain-on-snow melt events, and accelerating or decelerating snowmelt rates.

WATERMELON SNOW
Mixed media on canvas, 80x100 cm, 2019.

Mountain snowpacks often have a pink or reddish tinge in summer. This indicates the presence of algae which bloom near the snow surface. It has red pigmentation for protection against the high levels of ultra-violet radiation found at high altitudes in summer. This phenomenon is commonly known as watermelon snow. The red colouration makes the snow cover less reflective (typically around 15%) to radiation from the sun, causing faster snowmelt. Increasing temperatures due to global heating have produced earlier and more extensive high mountain snowmelt, and the resultant greater red algae growth, in turn, leads to faster and more extensive snowmelt - a positive feedback with negative consequences for mountain snowpacks and glaciers. Watermelon snow is evidence of how the snow ecosystem interacts with the atmosphere to change not only the colour of the mountains, but the downstream water supplies.

FORTRESS RIDGE STATION IN SUMMER
Field drawing, pastel on paper, 33x45 cm, 2019.

Global Water Futures scientists are trying to understand the fundamental processes that govern the interaction between climate, high mountains, snow, ecosystems and streamflow generation. They show that mountain hydrology is very sensitive to climate change and that ecosystem processes mediate this response leading to a different future ability of mountain catchments to generate source waters. This alpine environment is sensitive to drought and has recently suffered from very low soil moisture levels that have caused alpine plants to slow their photosynthesis and transpiration.

"The paintings I have produced of the Fortress Mountain Research Basin, I believe, are amongst the most dramatic I have produced. The weather was certainly dramatic: icy winds and blizzards on some occasions; practically balmy days on others. In summer, to my astonishment, rather than wind-hardened snow, the land around the automatic weather station closest to The Fortress peak had a green alpine meadow appearance. But, already, I could sense its return to its icy winter wilderness." - Ivanov

CANADIAN RIDGE STATION OVER THE KANANASKIS VALLEY

Oil on aluminium, 80x100 cm, 2021.

The Global Water Futures Observatories network in the Fortress Mountain Research Basin is extensive and high density, built to improve the understanding of rapid climate and environmental changes and to further develop predictive models of the effect of future climate change on the local hydrology – a particular challenge in such complex terrain. The Canadian Ridge station, a few hundred metres below the one on Fortress Ridge which has The Fortress as its sentinel, overlooks the deglaciated and now forested valley of the Kananaskis River.

THE FORTRESS PRESERVES A HIDDEN GLACIER

Oil on canvas, 30x30 cm, 2019.

The Fortress sustains a debris-covered glacier in the cirque valley at its base. Here, the remaining glacier ice slowly melts, feeding a small lake and complex groundwater flow systems. The debris-covered ice is still fed by windblown snow and snow avalanches and is shaded from solar radiation energy by the long shadow of The Fortress. It also receives rock avalanches which formed the layer of debris over the glacier ice. In some cases, glacier preservation is due to redistribution of the mountain snowpack and the insulation provided by a debris layer.

THE SLOW NUCLEAR EXPLOSION AT FORTRESS MOUNTAIN

Oil on canvas, 90x70 cm, 2019.

The natural cauldron beneath The Fortress, containing its hidden glacier, seemed a suitably dramatic location for Ivanov's impression of the Gwich'in Chief's "slow nuclear explosion" quoted in an article in the *Yukon News* on April 12, 2019. The Chief was describing the effect of global heating on his people's land of permafrost, further north in Northwest Territories and Yukon Territory, but both artist and scientists understood the universality of his words.

A POST-SCIENCE ERA?

Oil on canvas, 150x100 cm, 2020.

The slow nuclear explosion, unless we manage to contain it to paintings, will show that the decades-old warning of scientists has been ignored. In Ivanov's explosion the first thing to be destroyed is one of mountain science's most useful tools – a hydrometeorological observation station. One objective of the International Year for Glaciers' Preservation 2025 is to increase the monitoring of high mountain environments – always a challenge due to logistical difficulties, remoteness and cost. However, these stations provide immense benefits to downstream societies; from early warnings of extreme hydrological events, diagnostic understanding of mountain water and energy cycles, and improved model predictions of water supply. They show the changes occurring in high mountain environments – at least those short of the Gwich'in Chief's nuclear explosion.

MOUNT ROBSON
Field drawing, pastel on paper, 45x33 cm, 2022.

Ivanov's field painting captures the distinctive form of Mount Robson in British Columbia, Canada, as spring emerges in May after an exceptionally heavy snowfall season. The view is from the southwest; the Robson Glacier is on the eastern and northern side of the mountain, just out of sight. Mount Robson is the tallest peak in the Canadian Rockies and has the greatest vertical rise - almost 3 km - of any mountain in the Rocky Mountains of North America. Despite its high elevation and snowy environment, the Robson Glacier has retreated over 2 km since the early 1900s. It once fed rivers that flowed into both the Arctic and Pacific Ocean. However, glacier retreat means that now its glacial meltwater feeds only the Robson River, which is a tributary of the Fraser River which drains into the Pacific Ocean. The glacier is currently receding by more than 50 m per year. Rapid snowmelt in the record heat dome of June and July 2021 caused flooding that destroyed many of the trails to Berg Lake at the base of Mount Robson.

Mountains and Forests

WANING POWER
Field drawing, pastel on paper,
18x24 cm, 2019.

Victoria Glacier above Lake Louise in Banff National Park, Alberta, Canada, is a spectacular view in the Canadian Rocky Mountains UNESCO World Heritage Site. The glacier is an ever-diminishing shadow of its former self. On geological time-scales the immensity of its power is evident from the deep U-shaped valley, which it scoured out. Its vestigial form may vanish by the end of the century because of global heating. There seems to be a temporary contradiction in circumstances as people track towards the glacier over the still-frozen lake and snow in April.

MOROCCAN MEMORIES
Oil on canvas, 99x152 cm, 2024.

The snow-covered Atlas Mountains, perhaps, are reflecting on the short time - in their terms - when they were extensively glaciated in the Pleistocene. Africa has sustained glacier coverage in recent times in the Atlas Mountains as well as in East African mountains such as the Rwenzori, and Mounts Kenya and Kilimanjaro. Glacial moraines in the Atlas Mountains show that glaciers once extended to elevations as low as 2000 m. There are some signs that niche glaciers lasted into the 20th century and, even very recently, a small number of snow patches survived the summer season. This distant snowpack formed after a long snow drought and presages extreme mountain flooding that occurred from exceptionally heavy rainfall in 2024. North Africa's climate is becoming even more extreme as global heating proceeds.

AFRICAN SNOW – SOON TO BE LOST I, II

Field drawings, pastel on paper, 35x25 cm, 2024.

The Sun rises over the fore range of the Atlas Mountains in Morocco, and its heat will soon melt the light dusting of overnight snow. Mountain snowpacks provide the vast majority of streamflow runoff from, and groundwater recharge in, mountains around the world. This is threatened by rising temperatures which reduce the frequency of snowfall, increase that of rainfall, and advance the timing of snowmelt. Snowmelt rates can be increased by warmer conditions and greater solar radiation absorption by the snow as dust darkens the surface. The red Saharan dust does not remain in this environment but blows in the wind to the mountains of Spain, the Alps and Caucasus – adding a red veneer to the alpine snowpack and accelerating snowmelt.

Sub-alpine forest, Mount Idaho, British Columbia, Canada.

3.2 Forests

Sub-alpine, boreal and similar forests cover one fifth of the Earth's land area and are critical moderators of water sources in mountain and circumpolar regions, and act as regulators of the earth system. These forests are usually evergreen, needleleaf coniferous trees with shallow roots that are adapted to short cool summers and cold, snowy winters. Needleleaf forests have adapted to use water for photosynthesis sparingly and are known as nature's water managers due to their ability to slow snowmelt by shading it, then hold meltwaters in their deep organic soils and release runoff slowly through groundwater over the summer. Similar forests have covered the Earth for many millions of years. Needleleaf forests store vast amounts of carbon in their soils and associated peatlands and slowly absorb more carbon dioxide from the atmosphere than they release. But these forests also burn readily and have recently been subject to increasing disease and wildfires from climate heating and shorter snow seasons, as well as harvesting for wood products, pulp and paper. In Canada where the boreal forest dominates wooded land cover and most forests have long snowy winters, 62.6 million hectares of forest were lost from 2001 to 2024, equivalent to more than a 15% decrease in forest cover. In 2023 alone, more than 17.3 million hectares of land burned in Canada because of 5,475 forest fires. This was the largest annual loss of forests due to wildfires since records started. Despite the loss of many large continuous forests to wildfire, forests are slowly expanding at the alpine and arctic treelines in many parts of the world, covering what was previously open tundra with new, small, slow growing trees. These treeline forests collect windblown snow and darken the snow-covered landscape; this can accelerate permafrost thaw and global heating compared to the colder, snow-covered open tundra that they are replacing.

Yukon River near Whitehorse, Yukon Territory, Canada.

© Stacey Dumanski

Hydrometeorological tower in the Marmot Creek Research Basin, Alberta, Canada, and part of the Global Water Futures Observatories.

MARMOT CREEK RESEARCH BASIN, ALBERTA, CANADA
Oil on canvas, 20x20 cm, 2019.

This research basin has been instrumented since 1962 and is a key part of the Global Water Futures Observatories network. One of the original scientists was James Bruce who later co-proposed the Intergovernmental Panel on Climate Change (IPCC). Marmot Creek has been an important outdoor laboratory and crucible of scientific ideas and discoveries for dozens of scientists over more than 60 years. Forest covers immense areas of the circumpolar cold regions, and understanding forest energy and water balances is essential for developing regional and local predictive models. Needleleaf forests intercept a substantial portion of the seasonal snowfall and much of this sublimates back to the atmosphere. Such forests also slow snowmelt rates. The station's observations have helped in the understanding of climate change impacts on forests and have elucidated the role that forest disturbance plays in snow accumulation and snowmelt.

FOREST DEATH BENEATH STANLEY GLACIER
Field drawing, pastel on paper, 24x30 cm, 2019.

DEVASTATION BELOW STANLEY GLACIER
Field drawing, pastel on paper, 24x30 cm, 2019.

This field painting is of a scene close to the treeline beneath Stanley Glacier, British Columbia, Canada. Forests in North America and Eurasia have come under immense stress from climate change. Drought impedes growth and primes the undergrowth and needles for fire. Higher temperatures have meant that the pine bark beetle has been able to survive winters. Over-mature forest canopies - from decades of excessive artificial fire suppression measures - have also facilitated the survival and spread of the beetle. Die-back due to beetle infestation has provided tinderbox conditions for wildfires, the frequency and intensity of which has rocketed in recent years. Blackened trees are the evidence of fires; in this case from a burn in 2003. These forests take many years to regenerate.

This field painting of a recent forest burn gives an impression of the devastation caused by forest fires. The burn is in Kootenay National Park, part of the Canadian Rocky Mountains UNESCO World Heritage Site. In the last 20 years over 25% of this park has burned from wildfires caused by drought and lightning. Earlier snowmelt is causing an earlier and longer wildfire season and the area burned by wildfire has increased over time. In 2023, record wildfires across Canada burned 184,961 km², or about 5% of Canadian forest area, and resulted in catapulting Canada to becoming the fourth largest greenhouse gas emitter amongst nations by emitting 355 megatonnes of carbon to the atmosphere.

INFESTATION ALONG THE RIVER
Field drawing, pastel on paper, 30x24 cm, 2022.

Ivanov captures the impact of pine bark beetle infestation in killing the lodgepole pine forests along the banks of the Athabasca River, Jasper National Park, Alberta, Canada.

The pine beetle thrives in warmer winters and in intact forests – both of which prevail in Jasper Park due to human caused climate change and overmature pine canopies from decades of excessive wildfire suppression. This dead forest has since burned down in the 2024 Jasper wildfire.

© Gennadiy Ivanov

Damage to the underside of tree bark from the mountain pine beetle.

RED WARNING FOR MOUNTAIN FORESTS
Oil on canvas, 81x116 cm, 2022.

The mountain pine bark beetle damage is assumed to prime the forest for fire. But that may not always be the case; there are many factors at work. Modelling indicates that, because the trees remain standing for 3-5 years with their needles - which have turned a characteristic red - still on the branches, the forest is less productive and there is less detritus on the forest floor. After a few years, the needles fall off, and the forest turns grey; providing fuel for fire outbreaks. There are other complications, including moisture levels of the detritus and soil. These complications emphasise the importance of careful observations and the development of observation-informed predictive models.

The red near bankside portrays the waiting fire – which did come in 2024.

MOUNTAIN TINDERBOX AND ASHES
Oil on canvas, 90x70 cm, 2020.

Ivanov's evocative painting, with its red hues, shows the clear sign of pine bark beetle damage. Some entomologists call the beetles "first responders" to climate change, because they are so sensitive to it and the consequent environmental change. The beetles thrive when winter nights are less cold and in large mature swathes of forest. Winter daily low temperatures have increased by over 5 °C in this region since the 1960s providing ideal conditions for the beetle to survive the winter. The darker patches represent swathes of burnt trees, illustrating the importance of myriad factors, including local hydrology, soils and meteorology, controlling the patterns of damage and burn.

Slocan Valley, British Columbia, Canada.

FOREST FUTURES
Oil on canvas, 81x116 cm, 2022.

A kilometres-long lumber train carries healthy harvested wood through the Athabasca River Valley in Alberta, Canada. The red, dead, beetle-ridden, and burnt forest tracts seem to mock it from the other side of the valley.

VERMILLION PASS FOREST FIRE

Pastel on paper, 65x45 cm, 2019.

The location of this field painting is Stanley Creek, which is fed by melt from the Stanley Glacier, in British Columbia, Canada. Ivanov has let his imagine run riot over the colouration, inspired by the Pass's name - although the vermillion in the painting is very muted.

VERMILLION PASS BURN

Oil on canvas, 30x30 cm, 2019.

The artist has reverted to a colour scheme which more closely reflects the hues in the landscape in this oil painting of the Vermilion Pass burn.

OCHRE RIVER
Field drawing, pastel on paper, 24x30 cm, 2019.

Vermilion Pass gets its name from the colouration which originates in mineral springs of iron oxide. These ochre springs are important for the Indigenous people and are known as the paint pots. Scientifically, it is an interesting location because of the pronounced influence of groundwater chemistry on surface waters.

VERMILLION PASS BURN COLOUR
Oil on canvas, 80x80 cm, 2019.

This is an evocative painting which, unlike other evocative paintings, is tied to a specific location in Kootenay National Park, British Columbia, Canada.

"This felt a special – and elemental – place to me. The paint pots and the stream-water downstream were amongst the most vivid colourations I have seen in nature."
- Ivanov

SPRING NOW BRINGS DEATH, FIRE AND FLOODS
Oil on canvas, 160x160 cm, 2021.

Ivanov evokes what earlier and hotter springs bring - catastrophic floods and wildfires. Both 2023 and 2024 were record wildfire years in Canada – coincident with the hottest temperatures ever measured on Earth.

Canada, with one of the largest woodlands on the planet, lost 15% of its forested area in the last 24 years, much due to wildfire and disease associated with global heating. Burned soils and lack of needles in the needleleaf canopies mean that snowpacks are deeper and melt faster so that spring floods are flashier, more intense and destructive.

BURN!
Oil on canvas, 100x100 cm, 2019.

The painting evokes the earlier, and smaller snowmelt, combined with higher temperatures which prime the trees for burn, even in April, in northern and western Canada. The mountains behind the still-frozen lake fringed by its newly exposed silty desert are bereft of even their thin snow cover. A wisp of cloud resembles a glacier arm still clinging on. The burst of yellow and red on the mountains shout 'burn'!

FOX LAKE BURN
Mixed media on canvas, 90x120 cm, 2019.

In Yukon Territory, Canada between Whitehorse and Dawson City, there is a vast area (more than 45,000 hectares) of burnt forest, destroyed in a fire which raged in July 1998, and smouldered well into the following spring. The burn is still a scar, but - more than two decades on - small trees are now re-appearing.

FIRE AND ICE
Mixed media on canvas, 100x100 cm, 2023.

The albedo (reflectivity to solar radiation) on glacier surfaces can decrease from a summer average of about 0.29 before extensive wildfires to an average of 0.15 after extensive wildfires. The overall effect is complicated by the fact that smoky skies reduce solar radiation reaching the glacier surface but, on average, deposition of soot and ash onto the glacier can increase summer melt by 3% in the wildfire season and 10% in subsequent years.

THE WILDER FIRE
Oil on canvas, 100x100 cm, 2023.

Globally, forest wildfires are becoming more frequent and more damaging. In 2023 around 12 million hectares were destroyed, with Canada accounting for about two-thirds of the loss.

Post-glacial Valley in Rural Municipality of Enfield, Saskatchewan, Canada.

4. Downstream

Trevor Davies (left) and John Pomeroy (right) by Kluane Lake.

Climate-induced changes in the cryosphere are altering hydrology and water quality. Snow and ice are habitats for many species and are biologically active ecosystems. Anticipated ecological responses to a less snowy and rainier environment include increasing availability of liquid water near the surface throughout the year, increased vegetation at higher elevations and latitudes, enhanced nutrient and contaminant mobility, growth of algae and other microorganisms and greater organic carbon production. Water quality in warming cold regions is also of concern, with some studies suggesting that permafrost degradation may increase heavy metal concentrations and movement in groundwater. Impacts include changes in sediment and thermal regimes, shifts in biogeochemical and contaminant fluxes, changes in habitat availability and quality, and modifications in species biodiversity patterns. Glaciers play an important thermoregulating role in both freshwater and nearshore marine habitats, with glacier meltwater and groundwater emanating from ice-bearing debris-covered glaciers helping to maintain consistent, cool temperatures crucial for some fish species. Glaciers and snowfields can also sometimes be the main source of freshwater for alpine wetlands.

Water systems influenced by the cryosphere extend far downstream. Shifts in the timing of seasonal snowmelt and glacier melt and from relatively reliable melt to more variable and less predictable rainfall reduces water security. Snow and ice changes may affect downstream communities that are not obviously connected to mountain landscapes, making increased awareness of the cryosphere and its role in the global water cycle important, especially amongst water managers and other decision-makers. Management and infrastructure planning are often based on historical data, but the changing climate challenges the reliability of past patterns are predictors of the future. This is why enhancing observations and monitoring is so critical: we have designed substantial amounts of modern society to operate according to conditions that no longer exist and so adaptation to emerging hydroclimates is imperative and urgent.

4.1 Rivers and lakes

FIVE FINGER RAPIDS
Field drawing, pastel on paper, 30x42 cm, 2019.

The rapids are on the Yukon River, Canada, and were a tricky obstacle for the boats of the Klondike Gold Rush in the 1890s. The annual river ice break-up was an important event for such an important transport artery. The signs of early break-up can be seen in this field painting - the free water channel sweeping round in the left of the painting. The timing of break-up has been recorded upstream at Dawson City since 1896. The break-up was observed a few days after this scene was painted. It was the second-earliest on record.

TRANSITION
Oil on canvas, 81x116 cm, 2019.

This evocative painting portrays a retreating glacier, pock-marked by rock, mud and silt, overlooking a lake. Together they represent the transitions in the landscape and in the hydrological cycle: less ice, less snow, sometimes more water, sometimes less water. The title also gestures to the transitions which local communities must make in response to these changes; and to the transitions which all societies will have to make to mitigate the worst consequences of climate change, and to cope with the inevitable continuing changes.

MALIGNE LAKE SNOWPACK
Field drawing, pastel on paper, 30x24 cm, 2022.

Maligne Lake in Jasper National Park, Canada is another iconic view in the 23,600 km² Canadian Rockies UNESCO World Heritage Site. The lake's basin had a high snowpack and lake ice was blanketed by deep snow because of the late seasonal snowmelt in late May. Global heating is affecting precipitation patterns as the warmer air contains more water vapor. In high mountain regions this can sometimes mean more and later snow. Global heating can bring great inter-annual variability in weather patterns.

MALIGNE LAKE MELT
Field drawing, pastel on paper, 45x33 cm, 2022.

Maligne Lake in Jasper National Park, Alberta Canada, still has its winter ice cover and signs of coming melt are showing up. It is the largest natural lake in the Canadian Rockies and is fed by meltwater from extensive snowpacks and three glaciers.

LOWER KANANASKIS LAKE

Field drawing, pastel on paper, 45x33 cm, 2022.

This lake has also been turned into a reservoir. The level of the water is very low, exposing much of the lakebed, because of the late snowmelt at higher elevations. Later in the year the reservoir filled to near-record levels because of very high seasonal snowpacks.

MEDICINE LAKE

Field drawing, pastel on paper, 45x33 cm, 2022.

There were low water levels in Medicine Lake, Jasper National Park, Alberta, Canada, because of the late spring. Snowmelt from high elevations had not begun by late May. Snowmelt floods this valley, raising the water levels in this natural lake substantially in June as water backs up before re-entering the groundwater system. Its extreme fluctuations are due to complex groundwater flow systems and show how groundwater must be considered in prediction of mountain water resources.

UPPER KANANASKIS LAKE I, II
Field drawing, pastel on paper, 30x24 cm, 2022.

Upper Kananaskis Lake is a natural lake which has been turned into a reservoir. More extreme weather patterns caused by global heating has implications for water supply.

KINBASKET LAKE
Field drawing, pastel on paper, 45x33 cm, 2022.

This reservoir is in British Columbia, Canada, on the Columbia River system. It fills with snowmelt in most years but, this May, despite local snowmelt it remains very low and its dry lakebed is subject to dust storms - shown here in Ivanov's painting. The summer of 2022 was record dry in parts of British Columbia and low streamflows impacted hydroelectricity production, community water supply, and led to high salmon mortality.

PLUNDERED BY PIRACY
Field drawing, pastel on paper, 30x42 cm, 2019.

Large areas of what was lakebed as recently as 2016 are now exposed, and it is possible to walk over to what was an island on the biggest lake in the Yukon Territory, Kluane Lake, part of the 98,391 km² Kluane/Wrangell-St. Elias/Glacier Bay/Tatshenshini-Alsek UNESCO World Heritage Site which contains the world's largest non-polar icefield. Over the space of 4 days, the 150 m wide Slims River, which fed the lake, all but disappeared. The geomorphological process known as "river capture" or "river piracy" caused this. As the Kuskawulsh Glacier, which is fed by the Saint Elias Icefield, retreated because of climate change, its meltwater diverted down the Alsek river valley, starving the Slims River of much of its former supply. This is the first known example of river piracy caused by human-induced climate change. It was a dramatic event. The meltwater which once fed a river basin which drained west to the Bering Sea in 2016, now feeds a river basin which drains south to the Pacific Ocean.

THE DESERT ON THE LAKE
Oil on canvas, 90x150 cm, 2019.

A trickle – which remains of the river whose glacial meltwater source was pirated by another valley because of glacier recession – makes its contorted way towards Kluane Lake. Where once was lakewater, there is now silty dust. The strong cold katabatic winds, which flow off what remains of the Kaskawulsh Glacier, whip up dust storms. Melt from the thin snow cover on the distant mountains is not sufficient to maintain the lake at its pre-2016 level, when the act of piracy took place. Summer peak lake levels have been reduced by 1.6 m on average because of the meltwater grab. Global Water Futures projections to 2100 show that even with increased winter precipitation later this century, the lake levels will not come back. This sub-arctic 'desert' is here to stay.

DESTRUCTION BAY
Oil on canvas, 100x100 cm, 2019.

The dark, drying mud in the foreground is bounded by the bouldery walls of the now redundant Destruction Bay harbour, once used for the fishing boats of the Indigenous People for whom fishing on Kluane Lake is an important activity. The lake is still frozen – it is April - but the mountains in the background are free of snow. The useless harbour is on Destruction Bay: an apt name. It has since been rebuilt to accommodate the lower lake level.

THREE SISTERS OVERLOOK IMPENDING DROUGHT
Oil on canvas, 30x30 cm, 2019.

The foreground to the Three Sisters mountains near Canmore, Alberta, Canada, is a dry channel of the Bow River after a dry period of low flows in the summer. With reduced late summer glacier and snow melt inputs to the Bow River, it relies on groundwater discharge and rainfall to sustain streamflow. A long dry summer after a low snowmelt year, associated with global heating, like 2019, provides insufficient rainfall and groundwater to sustain streamflows. Groundwater depletion during mountain droughts is increasingly severe and is manifested in declining late summer, autumn and winter streamflows.

COUGAR CREEK DEBRIS NET
Field drawing, pastel on paper, 33x45 cm, 2019.

Cougar Creek is a tributary of the Bow River. On June 20, 2013, the town of Canmore, Alberta and much of the surrounding region experienced a devastating flood that was part of the most expensive natural disaster in Canadian history at that time. The torrent formed a debris flow that tore down Cougar Creek destroying homes, railways and roads, causing substantial damage and isolating the region for several days. The flood was caused by three days of heavy rainfall forming runoff over still frozen alpine soils and enhanced by melt of a late lying alpine snowpack. It was exceptional event in modern Canmore, but similar events were noted in the late 19th and early 20th centuries. In anticipation of the increased frequency and magnitude that is likely for future extreme rainfalls, a temporary debris net was constructed above the town after the 2013 flood to retain boulders and trees within the torrent. A retaining dam has now been built to replace this. The University of Saskatchewan's Coldwater Laboratory is based in Canmore, researching climate and hydrological regime changes and has developed predictive models for events such as this.

TRANSCONTINENTAL TRAIN BY BRÛLÉ LAKE
Field drawing, pastel on paper, 30x24 cm, 2022.

"As I sat painting this scene, one of the never-ending trains rumbled past. The steel tracks of the Rockies have witnessed dramatic changes in the landscape through which they were cut since the first trainlines were built in the late 1800s." - Ivanov

4.2 Ecosystem and Agriculture

SIBBALD PEATLAND I, II
Field drawings, pastel on paper, 30x24 cm, 2022.

Sibbald Fen in Kananaskis, Alberta, Canada, was instrumented in 2006 to study how beaver-created wetlands influence hydrology and water supply. In the massive flood of 2013, the beaver dams held and reduced flooding downstream in the city of Calgary. Studies of beaver-influenced hydrology show how ecosystem processes can restore hydrological stability even when extreme events, due to climate change, occur. Sibbald Fen is part of the Global Water Futures Observatories network of 64 instrumented research basins across Canada and was designatedas as a UNESCO Ecohydrology Demonstration Site in 2024.

Hundreds of miles to the east, the meltwater from the Canadian Rockies flows through the Canadian Prairies in the South Saskatchewan River. Streamflow regimes in the Canadian Prairies are also changing because of climate change, wetland drainage and human use for irrigation, communities and hydroelectric power.

SHORT SOJOURN ON SOUTH SASKATCHEWAN
Field drawing, pastel on paper, 24x30 cm, 2019.

Pronounced floods and droughts (which are leading to earlier and more frequent vegetation fires) are increasing in frequency and intensity, with implications for agriculture, infrastructure and transport. The worst floods and droughts since colonisation of the region in the late 1800s have occurred since 2000. The South Saskatchewan River flows through the city of Saskatoon, the home of the University of Saskatchewan and the headquarters of the Global Water Futures programme. This view of Cranberry Flats, a series of sand dunes overlooking the river, shows slumping riverbanks and the changing pattern of sand bars. This is strong evidence of shifting patterns of erosion and deposition in response to recent hydrological changes. The best way of seeing this evidence is from a canoe, and this pastel illustrates the changing views – very different from the sculpted landscapes of the Rockies – painted on a 12 km sojourn on the river, downstream towards Saskatoon, under the paddle power of the scientists!

NORTH EAST SWALE
Field drawing, pastel on paper, 18x24 cm, 2019.

North East Swale is on the edge of Saskatoon, Saskatchewan, Canada. It is a unique environment, which has natural and cultural connections to Saskatchewan's past. It is a habitat for more than 200 plants species, more than 100 birds, and many mammals, reptiles, insects and amphibians. It harbours some of the last Plains Rough Fescue in the world. Its wetlands store runoff from the rapidly growing immediately-adjacent urban neighbourhood. Consequently, its successful management is challenging. It is an important field training site for students at the University of Saskatchewan. Here they start to understand the rigorous science methods which are required to understand the complex human-environment linkages which are needed for sustainable management of valuable environmental resources. This field pastel shows a new housing development in the distance; and, behind it, an approaching storm. There are dead patches in the vegetation cover – a consequence of pronounced fluctuations in water levels due to recent drought and previous flooding extremes that are magnified because of climate change.

CLAVET RESEARCH FARM, SASKATCHEWAN, CANADA
Field drawing, pastel on paper, 24x30 cm, 2019.

Just outside Saskatoon is a Global Water Futures Observatories agricultural water research site at the Clavet Livestock and Forage Centre of Excellence: a University of Saskatchewan research farm. Ground-based instrumentation and drone-borne sensors help scientists understand, and model, hydrological processes in this intensive agricultural landscape. The linkage between retaining windblown snow on fields in winter, suppressing evaporation in spring by leaving plant residue on the fields, and crop production was shown here. Besides contributing to more effective agricultural practices – important as climate change leads to changing weather patterns – the data are important for modelling exchanges between crops and atmosphere and for understanding and predicting prairie agro-hydrology.

THE SALT LAKE
Oil on canvas, 80x80 cm, 2019.

Near Clavet Research Farm, the changing patterns of flood, salinisation and drought have led to the evaporation of a shallow lake, leaving salt deposits. In the distance are hay bales and a traditional grain elevator. Ivanov has used this scene to produce another evocative painting. He has supplemented the hay bales with stacks of oil pipe sections, which are a common sight across the Prairies. He has also introduced a coal-fired power station. Around one-third of Saskatchewan's power comes from coal-burning, from which some of the emitted carbon dioxide is sequestered underground.

PERMAFROST THAWING RELEASES FOSSIL FUEL

Oil on canvas, 70x90 cm, 2020.

The extraction and use of fossil fuel world-wide has led to a catastrophic thawing of permafrost. It is ironic that permafrost thaw can lead to the release of contained fossil fuel into the environment. This painting is based on satellite imagery of a major pollution incident at Norilsk in Siberia, about 500 km from the Kara Sea.

The incident started on 29 May 2020. Around 17,500 tonnes of diesel stored at a power plant leaked into the local waterways turning them red. The Kremlin declared a state of emergency. A major cause was permafrost thawing which caused the collapse of the storage tank. On 4 June, booms on the Ambarnaya River, which has been installed to contain the spill, were broken by large amounts of drifting ice. After the incident, which followed weeks of abnormally warm weather, the Russian authorities ordered urgent safety checks on all potentially hazardous installations built on Arctic permafrost.

The tundra polygon cracks visible in this painting portray the shifting ice-laden soils of Arctic permafrost which have become more dynamic and unstable as deeper and deeper thaw sets in. This puts at risk not only Arctic rivers but the massive infrastructure in northern latitudes such as mines, pipelines, roads, railways and communities. The disaster at Norilsk illustrates a common problem that circumpolar countries have in sustaining northern economies and environmental protection during the climate crisis.

4.3 Predictions

DENIERS' STATION
Oil on canvas, 95x90 cm, 2019.

Ivanov's inspiration for this painting was the cold regions field research stations of the Global Water Future Observatories programme. The instruments record meteorological variables including water vapour flux and carbon dioxide exchange. Ivanov has inverted the instruments since, to him, it represents the way in which climate deniers turn logic upside-down. Astonishingly, given the available evidence, some - including influential figures - still proclaim that human-induced climate change is a hoax. There are other forms of denialism. One is that climate change caused by humans is real but that it is just too difficult and expensive to do anything about it. It is difficult, but it will be very much more expensive not to do anything about it and, by and large, the solutions already exist. Another variation is that new technology will come galloping to the rescue. One such technology is carbon capture and storage but there is no indication that it can be deployed at the required scale or on the required timescale. Very sharp emissions reductions are required.

ARCTIC HEATWAVE
Oil on canvas, 90x116 cm, 2023.

There is considerable uncertainty around climate predictions for the Arctic regions, but it is expected that the Arctic will continue to heat faster than other parts of the planet. Summer sea ice in the Arctic could be relatively uncommon by 2050; it possible that the Arctic will be free of ice in winter months in the second half of this century unless we make severe cuts in global heating gas emissions. More precipitation is expected with a transition to rainfall-dominated precipitation in the summer and autumn. Permafrost thaw will continue at a rapid pace, releasing stored methane, a potent greenhouse gas, from thawed soils to the atmosphere. These changes will have profound impacts on Arctic ecosystems, global ocean and atmospheric circulation, and socio-economic activity.

IMPERMANENT FROST
Oil on canvas, 135x170 cm, 2019.

Permafrost is defined as a subsurface layer of soil that remains frozen throughout the year or, in other words, is "permanent". There is an enormous area of permafrost across northern Canada. Because of climate change it is no longer permanent. Large swathes of permafrost are thawing. Global Water Futures scientists have calculated that 90% of the permafrost in the Yukon and Mackenzie river basins will have thawed by the 2090s under a business-as-usual climate scenario. The consequences are dramatic. The surface layers slide down slopes, sometimes in slow viscous rivers of mud, carrying any vegetation, including trees, with them and leaving bare scars. When the permafrost is relatively water-rich, ice within the soil is clearly visible in the exposed scar, and during warm weather, water from the melting ice cascades out of the exposed soil. As the permafrost thaws, wetlands expand, forests collapse and die off, and vast quantities of carbon currently locked within the ground ice are released as greenhouse gases. For circumpolar permafrost, the 21st century is truly a time of the 'great thaw'.

SAVE THE SASKATCHEWAN PRAIRIES!

Pastel on paper, 45x64 cm, 2019.

As the Transitions team travelled eastwards from the Rockies, over the Alberta – Saskatchewan border, on the way to Saskatoon where the Global Water Futures programme has its headquarters at the University of Saskatchewan, the railways became an increasingly obvious feature in the landscape. They are used to transport Prairie wheat, oilseeds, pulse crops, potash and oil. Saskatchewan has the world's largest potash reserves, with many of the mines near Saskatoon. The resource formed when a large inland sea evaporated around 400 million years ago. The oil and gas industry is smaller than in Alberta but is still substantial: Saskatchewan is Canada's second largest producer of oil and has one of the highest greenhouse gas emission rates per person in the world. The traditional, tall, grain elevators by the rail-side are noticeable features, often greatly-valued as heritage by the settlements near where they are located. The STOP sign in this expressionistic painting signals humanity's need to stop just carrying on as "normal". If we do that the fertile breadbasket of the Canadian Prairies can still be saved.

BASKET CASE
Oil on canvas, 81x16 cm, 2023.

A painting of Kinbasket Lake has already appeared, earlier in the book. It is a 216 km long reservoir which was formed by the building of the Mica Dam in 1973, and its purpose is to help control the flow to hydroelectric stations lower down the Columbia River basin. This oil painting emphasises the dramatic hydrological changes in recent decades; to precipitation and snowmelt regimes and with increasingly severe droughts and floods. The dry reservoir bed is disturbed by a dust devil and the prolonged dry conditions are epitomised by the dry driftwood in the foreground; redolent of the skeleton of a steer. Highly variable weather, with increasingly severe and more frequent extremes, will disrupt the original plans for this reservoir, as it will for many other engineering projects dependent on stationary mountain hydrology.

Forest cultivation, Lo Manthang, Nepal.

4.4 Challenges and Solutions

"Many of the fabulous fossils in the Tyrell Museum just outside Drumheller evoked a horned serpent in my mind."
- Ivanov

THE DINOSAURS' DAMNATION
Oil on canvas, 120x150 cm, 2019.

Drumheller, Alberta, is an old coal-mining town. It is located on the Red Deer River which is now fed by mountain snowmelt and a small glacier contribution. As the continental icesheet melted at the end of the ice age, floodwaters carved out a spectacular canyon which is now occupied by the Red Deer River. The canyon walls have yielded globally important dinosaur fossils. Ivanov has taken the opportunity to juxtapose a dinosaur, the fossil- and coal-bearing strata, and a modern Alberta oil extraction facility. Many Indigenous peoples in North America have a mythology about a horned serpent, with the mystical beast associated with extreme events – rain, water, thunder, lightning. The dinosaur is watching, atop the beds overlaying the fossil-rich strata, as humankind is instigating a global environmental change which may well appear in Earth's future geological record - and is already dubbed the Anthropocene by some scientists.

LISTEN TO THE SCIENTISTS
Oil on canvas, 200x140 cm, 2020.

The title of this painting is a phrase which is universally associated with climate activist, Greta Thunberg. She visited the Global Water Futures study site on the Athabasca Glacier during a snowstorm in early winter 2019. Pomeroy discussed research results with her, including the effect of wind on snow dynamics on the ice. Ivanov has included one of his favourite icons in his work on climate change.

5. Conclusions

Destruction of Happisburgh beach, Norfolk, United Kingdom.

Scientists have been warning about the impacts of climate change for decades. In 1988 the Intergovernmental Panel on Climate Change (IPCC) was established to provide regular assessments of climate science and to make recommendations to policy makers. Over the years, thousands of scientists have been involved. IPCC Reports have been lauded for the thoroughness and high quality of their assessments and recommendations. The IPCC has been a remarkably successful undertaking in respect of bringing together the best of climate science and offering that science to policy makers.

Each year, since 1995 (except for 2020), a Conference of the Parties (COP) has been convened to bring together States which are Parties to the United Nations Framework Convention on Climate Change (UNFCCC). The objective of the Convention is to stabilise atmospheric greenhouse gas concentrations at a level which will avoid dangerous climate change.

Despite these major international efforts, and some regional declines in emissions, global carbon dioxide emissions have increased by more than 1.5 times since 1990. The funds committed to loss and damage from climate change are far less than actual costs. In this regard, further efforts are needed to support decision and policy makers to make climate-informed decisions and support the adoption of resilience enhancing strategies. Denialism of climate science, observations and model predictions for climate change have long been around, sometimes rearing its head in spectacular form, such as in "Climategate" in 2009-2010. Nevertheless, the science has proven credible, and predictions of models successfully validated for the period of observation. Despite this, the counterfactual narratives have had deleterious impacts on public policy and puts large segments of the Earth's ecosystems and human population at considerable risk. The inevitable conclusion – that fossil fuel burning needs to be reduced - was recognized at COP28 in 2023 which reached an agreement for the "beginning of the end" of the fossil fuel era by laying the ground for a swift, just and equitable transition, underpinned by deep emissions cuts and scaled-up finance. It is now up to the world to meet these aspirations and to preserve our glaciers, snow, permafrost and therefore ourselves.

The impacts of climate change, via heatwaves, storms, floods, droughts, and wildfires, have hit most communities in some form. There are likely to be more pervasive impacts, such as sea-level rise to an extent which will directly impact hundreds of millions of people, or major changes in the world's ocean currents which will lead to large and long-lasting disruptions to global weather systems such as hurricanes, monsoons and the North Atlantic circulations that currently keep Europe temperate.

Many early human societies and most current Indigenous cultures understood from direct experience that they were part of, and had to live in balance with, their environment. They had, or have, a connection with their surroundings at an emotional level, a respect for their environment and local ecosystems. Urbanisation and intensification of global trade is one of the of the reasons why so many people no longer have this connection to their local and regional environments and resources.

Culture and the arts are powerful resources and tools in our collective pursuit of a sustainable future. They mobilize global communities, shape public narratives, and drive transformative policy action on climate change. Ultimately, by harnessing compelling storytelling and historical insights, culture not only shapes the public narrative around climate change but also offers communities a means to cope with the profound anxiety and loss that can accompany environmental transformation. Culture has the power to foster empathy and deeper connection to environmental issues, while challenging perspectives and initiating dialogue in order to mobilize action and drive lasting change.

One thing that has struck the scientists in the Transitions team, at talks and in exhibitions, is how many people say that they have an emotional reaction to Ivanov's paintings of the impacts of climate change. The science-informed art seems to stimulate a connection of people to the state, or fate, of our environment. Perhaps art needs to be used more widely when attempts are made to engage people in the fate of our world. After all, it is pressure from the public (in most states) which persuades politicians that they should shift priorities.

A recent survey across 125 countries found that 89% of respondents "demand(ed) intensified political action" on climate change, and 69% were willing to contribute 1% of their personal income. The authors of the study conclude that individuals systematically underestimate the willingness of others to act. This is a tragic gap. There is latent, but not yet released, pressure to be exerted on politicians. Tried catalysts have not yet worked. A suite of catalysts is needed. Art should be in there.

We live in an age where the cryosphere is being destroyed by excessive heat, where the cold regions of the world are warming at faster rates than elsewhere, and where downstream water supply is becoming less reliable and more erratic – fluctuating between droughts and floods. This is a dangerous age, one where the cryospheric buffers that have provided water resources for billions of people cannot be relied upon much longer, where ancient components of our mountain landscapes such as glaciers and icefields are crumbling before our eyes, where the vast northern forests are collapsing along with their formerly permafrost-laden soils, and where wildfire has become an annual horror of the circumpolar summer. The burning of much of Jasper National Park UNESCO World Heritage Site and one-third of the town of Jasper, Canada, in July 2024 has shown how transient the current cold regions landscape is and how endangered our homes are - in every sense of the word.

Collaboration and cooperation between GWF and UNESCO has given GWF's Transitions team an opportunity to join Ivanov's art to the fate of glaciers, snow and ice - those emblematic canaries, or . indicators, of the state of our climate and hydrology. The art was exhibited and then the topic discussed as part of an event organized with UNESCO-IHP at COP-26 in Glasgow and then shown at UNESCO-IHP events to lead discussions at COP-28 and COP-29. It is merely a book, and so the textures, and the scale of some of the paintings are missing. But it is hoped that a few more people, after seeing photographs of Ivanov's paintings, will start to experience the emotional connection to our stressed environment which so many, around the world, have lost. Not all, of course. Those who are most immediately in danger, from rising sea level, glacier lake outburst floods, wildfires, declining ice cover, or thawing permafrost, for example, don't need art to raise their awareness.

Further, this book documents the state of glaciers, snow, permafrost, forests, lakes, rivers and wetlands as they were in 2019-2022 in northwestern Canada, including several UNESCO World Heritage sites. This will be invaluable for future generations so that they remember the beauty and grandeur of these landscapes and of the hydrological and earth system function that the cryosphere provided to us all. What does a post-cryosphere future hold for us? What will our ecosystems be composed of and how will they function? Will there be enough clean freshwater to drink and to produce food and energy? Can we survive in such an alien world? The answers to these questions are uncertain and there is a possibility, perhaps a likelihood, that this post-cryospheric future will have poor prospects for humanity and much of nature. There is still time to turn to a different future and to avert the worst losses of the cryosphere. But the challenges to comply with agreed commitments to reduce greenhouse gas emissions, and the recent rise in temperatures to more than 1.5 °C greater than the pre-industrial normal range tells us that there is very little time left. Action will have to be fast and profound. We hope that this book helps impel that action.

THE END OF THE ENDLESS OIL TRAIN
Oil on canvas, 90x90 cm, 2019.

Waiting at railway crossings, the oil trains traversing the Canadian Prairies seemed endless. Many were transporting oil, from the Athabasca oil sands and conventional oil formations in the region, to refineries and markets far away. Eventually they did end. Is this painting prophetic?

Transporting oil this way presents risks from spills and contamination. Two derailments in two months in 2019 near Guernsey, 115 km south-east of Saskatoon, Saskatchewan, led to spills of 1.5 and 1.6 million litres of oil. These incidents can be catastrophic. The derailment of a train carrying western oil in 2013 caused an explosion and the deaths of 47 people in Lac Mégantic, Québec, Canada.

SADNESS AT HAPPISBURGH

Oil on canvas, 109x81 cm, 2018.

In Norfolk, England, where Ivanov and Davies are based, the impacts of rising sea level due to glacier melt and ocean temperature increases, and increasing storms due to a changing climate, are apparent in this receding coastline where ancient villages are being destroyed. Homes are falling into the sea and defending all of them against this destructive global heating trifecta of melting glaciers, warming oceans and increasing coastal storms is, sadly, impossible. The fate of small island nations is especially sad and whole chains of inhabited islands are expected to be covered by rising seas in this century. This will remove the land base for some nation states.

"Since I have been working with scientists, I have started to understand how humans are affecting the whole world, and how impacts in one region have consequences for other regions. Meltwater from glaciers, wherever they are, is contributing to increasing coastal erosion very close to home".
- Ivanov

INUNDATION
Oil on hessian, 81x109 cm, 2018.

Norfolk, England has some of the most rapidly eroding coastline in Europe. Coastal erosion at Happisburgh is where recent coastal erosion has been most severe in Norfolk. Around 40 Happisburgh homes have tumbled into the sea in the last 20 years. Rising sea-level, due to global warming, is exacerbating the problems which the erodible coastline faces. Most of the erosion occurs during dramatic events associated with storms. The impact on the local community is great. This painting was produced after a severe episode in 2018, when there was pronounced cliff collapse and damage to the timber sea defences. Predictions indicate more future sadness from coastal retreat.

THE REVENANT
Oil on canvas, 80x100 cm, 2023.

The title refers to published scientific predictions indicating that much of Dubai's infrastructure could be underwater by 2100, mostly as the result of mountain glacier and polar ice sheet melting. These are issues that everyone should care about. Pomeroy showed this, and other Transitions paintings, at the 2023 United Nations Climate Change Conference (COP28) in Dubai. Ivanov paints the city of Dubai being overwhelmed by sea water. Over it lurks a faint image of a revenant Peyto Glacier. Peyto Glacier is not expected to last through this decade. As Peyto Glacier melts away, coastal cities such as Dubai will be subject to more frequent flooding and eventual inundation due to the law of mass conservation. The mountains and oceans are intimately connected by freshwater.

CASPIAN OIL FUTURES
Oil on canvas, 80x100 cm, 2024.

The Caspian Basin is an important location for oil and gas extraction. The landlocked Caspian Sea depends partly on the melting of snowpacks in the adjoining Caucasus Mountains for recharge. These mountains are vulnerable to changes in snowpacks and streamflow, and the Caspian Sea level is projected to drop as much as 60 m by the end of this century due to reduced streamflow inputs and rising evaporation rates. Here, Ivanov depicts a future dystopian future with wildfires in the dry mountains, and a dry Caspian sea-bed exposing only pipelines and drilling rigs.

ICE AND FIRE

Oil on canvas, 116x91 cm. 2019.

Here the artist expresses some of his memories of the Transitions trip to the Rockies and the Prairies. One is the seemingly endless trains which transport Canadian resources for export and carry manufactured goods from around the globe – a symbol of the nature of human activity which is driving climate change. Also included is a Global Water Futures' observation station which monitors the changing weather due to climate change, and the changing state of glaciers, snowpacks and water supply. A scientist reflects on the challenges global society faces in decoupling our necessary activities from continuing increases in global heating gas emissions. There is no time for reflection. We need the widespread adoption of existing low carbon technologies and a rapid re-ordering of how we go about our lives and business. We need to develop solutions to the global water crisis to reduce the worst impacts of the global climate crisis. There is much to do.

MARCHING TREES
Oil on canvas, 90x100 cm, 2019.

With higher summer temperatures, trees are growing further northward and at higher elevations. They colonise firm land which was previously tundra. First, one or two individuals establish themselves, followed by more until there are many scattered individuals or, in favoured locations, small krumholtz. This evocative painting recalls many landscape elements in the Ogilvie Mountains of the northern Yukon Territory, Canada, especially where the terrain has channeled blowing snowstorms, causing the trees to contort and bend. They looked as though they were, slowly, but inexorably, marching northwards. As the forest expands, the trees restrain snow from blowing, heat in the spring sun and help to thaw frozen ground. Major ecosystem changes are afoot.

Wolf Creek Research Basin, Yukon Territory, Canada, and part of the Global Water Futures Observatories.

THE CLOCK IS TICKING
Oil on canvas, 150x150 cm, 2023.

The presence of the clock needs no explanation. Ivanov has chosen to paint it into the background of a beaver and its dam, the subject of a study by Global Water Futures (GWF) into beaver ecohydrology. The beaver was almost trapped into local extirpation by the fur trade in the early 20th century. Beaver ponds provide ecological services in helping to modulate downstream floods and promote groundwater recharge. They also increase methane (a powerful global heating gas) emissions because of the trapping of organic matter in their ponds. GWF researchers have concluded that beavers are of considerable net benefit, but this is a handy reminder that environmental processes are complicated, and political decisions need to be informed by the best possible science.

SUNSET ON COLD REGIONS
Field drawing, pastel on paper, 30x42 cm, 2019.

The mountains near Haines Junction, Yukon Territory, are hauntingly attractive at sunset. The single spruce tree portends mourning for a lost past and apprehension of an uncertain future. We must accept that, with the emissions to date of long-lived greenhouse gases, we are committed to a substantial degree of global warming. This will inevitably mean that many glaciers will melt, much permafrost will thaw, snow seasons will shorten and snowpacks will decline in the next decades. We will not preserve all glaciers, and this will mean the termination of many cryospheric components of our earth system and the emergence of new, warmer, more rainfall-dominated river basins, expanded vegetation cover in cold regions, migration of species, and massive adaptation needs for humans. The world will have greatly diminished cold regions and less reliable downstream water supplies. This will further challenge our attempts to achieve freshwater sustainability by 2030. We must face this challenge and keep going to both reduce greenhouse gas emission and manage freshwater in innovative ways so that we can create our own global water futures and recreate a world of beauty, decency, diversity and prosperity.

CRYOSPHERE SUNRISE
Oil on canvas, 100x100 cm, 2019.

This painting of sunrise on the Front Ranges near Haines Junction, Yukon Territory raises the question - "What sort of world will emerge in the future as climate change proceeds, the cryosphere shrinks and water supplies, ecosystems and humanity respond and adapt?" The answer to this question is partly for the natural sciences to answer – but incompletely. The social sciences must help us understand how individuals and communities will adapt and respond to the impacts of climate change, and how nations and the international community will respond to the need to reduce greenhouse gas emissions to avoid the most damaging consequences of climate change. There will still be inevitable unavoidable loss and damage which we will have to address. We will have to reach deep into our own humanity, and include traditional understandings from diverse cultures, to find equitable solutions. At times one can lose hope in the future, but we must remember how resilient life on Earth is, how much knowledge our societies hold, how brilliant and capable our youth are, and how proficient humans are as water managers. We have repeatedly demonstrated the ability of our species to work together, to rise above expectations and to "beat the odds" by addressing massive challenges. Humanity is part of the earth system, and we will shape how it develops. We can shape this transition towards sustainability. The time for this transition is now and we hope that this book encourages those who would transform our world for this new day.

6. Suggested Reading

- Andre P., Boneva, T., Chopra, F., & Falk, A. 2024. Globally representative evidence on the actual and perceived support for climate action. *Nature Climate Change* 14, 253-259 https://doi.org/10.1038/s41558-024-01925-3

- Aubry-Wake, C., Bertoncini, A. and Pomeroy, J. W. 2022. Fire and ice: The impact of wildfire-affected albedo and irradiance on glacier melt. *Earth's Future*, Vol. 10, No. 4, Article e2022EF002685. https://doi.org/10.1029/2022EF002685

- Barnett, T. P., Adam, J. C. and Lettenmaier, D. P. 2005. Potential impacts of a warming climate on water availability in snow-dominated regions. *Nature*, Vol. 438, No. 7066, pp. 303–309. https://doi.org/10.1038/nature04141.

- Fang, X. and Pomeroy J.W. 2023. Simulation of the impact of future changes in climate on the hydrology of Bow River headwater basins in the Canadian Rockies. *Journal of Hydrology*, Vol. 620, pp.129566, https://doi.org/10.1016/j.jhydrol.2023.129566.

- Hugonnet, R., McNabb, R., Berthier, E., Menounos, B., Nuth, C., Girod, L., Farinotti., D., Huss, M., Dussaillant, I., Brun, F. and Kääb, A. 2021. Accelerated global glacier mass loss in the early twenty-first century. *Nature*, Vol. 592, pp. 726–731. https://doi.org/10.1038/s41586-021-03436-z.

- Huss, M. and Hock, R. 2018. Global-scale hydrological response to future glacier mass loss. *Nature Climate Change*, Vol. 8, No. 2, pp. 135–140. https://doi.org/10.1038/s41558-017-0049-x.

- Immerzeel, W. W., Lutz, A. F., Andrade, M., Bahl, A., Biemans, H., Bolch, T., Hyde, Brumby, S., Davies, B. J., Elmore, A. C., Emmer, A., Feng, M., Fernández, A., Haritashya, U., Kargel, J. S., Koppes, M., Kraaijenbrink, P. D. A., Kulkarni, A. V., Mayewski, P. A., Nepal, S., Pacheco, P., Painter, T. H., Pellicciotti, F., Rajaram, H., Rupper, S., Sinisalo, A., Shrestha, A. B., Viviroli, D., Wada, W., Xiao, C., Yao, T. and Baillie, J. E. M. 2020. Importance and vulnerability of the world's water towers. *Nature*, Vol. 577, pp. 364–369. https://doi.org/10.1038/s41586-019-1822-y.

Suggested Reading

- Intergovernmental Panel on Climate Change, Climate Change 2023, Synthesis Report https://www.ipcc.ch/report/ar6/syr/

- Jones, H. G., Pomeroy, J. W., Walker, D. A. and Hoham, R. W. (eds). 2001. Snow Ecology: An Interdisciplinary Examination of Snow-Covered Ecosystems. Cambridge, UK, Cambridge University Press. https://doi.org/10.1111/j.1365-2745.2001.610-3.x.

- López-Moreno, J. I., Pomeroy, J. W., Alonso-González, E., Morán-Tejeda, E., and Revuelto, J. 2020. Decoupling of warming mountain snowpacks from hydrological regimes. *Environmental Research Letters*, 15(11), 114006. https://doi.org/10.1088/1748-9326/abb55f

- Pepin, N. C., Arnone, E., Gobiet, A., Haslinger, K., Kotlarski, S., Notarnicola, C., Palazzi, E., Seibert, P., Serafin, S., Schöner, W., Terzago, S., Thornton, J. M., Vuille, M. and Adler, C. 2022. Climate changes and their elevational patterns in the mountains of the world. *Reviews of Geophysics*, Vol. 60, No. 1, Article e2020RG000730. https://doi.org/10.1029/2020RG000730.

- Rounce, D. R., Hock, R., Maussion, F., Hugonnet, R., Kochtitzky, W., Huss, M., Berthier, E., Compagno, L., Copland, L., Farinotti, D., Menounos, B. and McNabb, R. W. 2023. Global glacier change in the 21st century: Every increase in temperature matters. Science, Vol. 379, No. 6627, pp. 78–83. https://doi.org/10.1126/science.abo1324.

Yakou Snow Observation Site at 4,120 m elevation in the Qilian Mountains, China.

References

Jouzel, J., Masson-Delmotte, V., Cattani, O., Dreyfus, G., Falourd, S., Hoffmann, G., Minster, B., Nouet, J., Barnola, J., Chappellaz, J. (2007). EPICA Dome C ice core 800kyr deuterium data and temperature estimates. *IGBP PAGES/World Data Center for Paleoclimatology Data Contribution Series*, 91, 2007. https://doi.org/10.1126/science.1141038.

LisiecCki, L. E., & Raymo, M. E. (2005). Correction to "A Pliocene-Pleistocene stack of 57 globally distributed benthic $\delta^{18}O$ records": CORRECTION. *Paleoceanography*, 20(2), n/a-n/a. https://doi.org/10.1029/2005PA001164

Marcott, S. A., Shakun, J. D., Clark, P. U., & Mix, A. C. (2013). A Reconstruction of Regional and Global Temperature for the Past 11,300 Years. *Science*, 339(6124), 1198–1201. https://doi.org/10.1126/science.1228026

PAGES 2k Consortium. (2019). Consistent multidecadal variability in global temperature reconstructions and simulations over the Common Era. *Nature Geoscience*, 12, 643-649. https://doi.org/10.1038/s41561-019-0400-0

Zachos, J. C., Dickens, G. R., & Zeebe, R. E. (2008). An early Cenozoic perspective on greenhouse warming and carbon-cycle dynamics. *Nature*, 451(7176), 279–283. https://doi.org/10.1038/nature06588

John Pomeroy (top left) and Trevor Davies (top right) on Peyto Glacier as Gennadiy Ivanov (bottom) plein air paints with pastels.

© Mark Ferguson